LUSITANIA

LUSITANIA

AN ILLUSTRATED BIOGRAPHY (VOLUME ONE)

LIFE OF A GREYHOUND

J. KENT LAYTON, TAD FITCH, MICHAEL POIRIER,
TOM LYNSKEY, LEVI ROURKE

Cover illustrations courtesy of HFX Studios.

Previously published in earlier forms as *Lusitania: An Illustrated Biography of the Ship of Splendor* (Lulu.com: 2007) and *Lusitania: An Illustrated Biography* (Amberley: 2015).

This edition first published 2024

The History Press
97 St George's Place
Cheltenham
GL50 3QB

www.thehistorypress.co.uk

British Library Cataloguing in Publication Data.
A catalogue record for this book is available from the British Library.

ISBN 978 1 80399 523 6

Typesetting and origination by The History Press
Printed by Thomson Press, India

In memory of one of the greatest ships the world has ever known, which reliably delivered hundreds of thousands of passengers safely across the Atlantic for over seven years: The Greyhound *Lusitania*.

J. Kent Layton: For my wife, Tessa Anne.

Tad Fitch: For Jocelyn, Jackie, and my father.

Michael Poirier: For my late friend Barbara Anderson McDermott, the last *Lusitania* survivor with memories of the voyage.

Tom Lynskey: For my wife, Emma, my parents, and my little dude, Tommy.

Levi Rourke: For my parents, siblings, and friends.

Frontispiece:
Lusitania sailing east towards home at dusk. (HFX Studios)

Left: Lusitania in the Canada Graving Dock before her formal trials. (HFX Studios)

CONTENTS

FOREWORD
BY STUART WILLIAMSON

The *Lusitania* has been my passion from my early teenage years, so to be asked to write this foreword by such incredible historians was a complete honour and privilege.

My interest in *Lusitania* began with watching intently the *Newsnight* programmes in 1982 about the salvage operations that were being conducted by Oceaneering at the time; it drove a passion in me to learn more about the wreck, and in doing so I've spent many years researching the history of dives to the wreck site and became involved with *Lusitania* Project 17, a dive team lead by Peter McCamley dedicated to mapping, learning more and preserving the legacy of the great ship.

In my time as a professional artist, I did my best to bring *Lusitania* back to life. Over the last six years with dive team Project 17, as their historian making a detailed study of the wreck I have tried to imagine the Grand Lady as she once was. But nothing brings to life the beauty and the human story of that beautiful ship more than the pages of this book.

History, the fog of war, and time have in some ways overshadowed the history of *Lusitania*, but when she was built, she was the pinnacle of her age; Britain's first turbine-driven, four-funnelled liner, and that excitement and feeling of pride can be summed up by descriptions of her maiden departure from Liverpool, the world's largest liner at that time:

> Probably never before has so much interest been displayed in the maiden voyage of a new vessel as was displayed today in the sailing of the *Lusitania*. This can be attributed partly to the intense rivalry between England and Germany in the transatlantic passenger traffic, partly to national pride in ownership – she is believed to be the greatest triumph of the shipbuilder's art afloat – and to no small extent to the sporting element given to the event by the prospect of a race between the youngest Atlantic flier and the *Lucania*, at one time the holder of the record.
>
> The *Lusitania* was over one hour late in starting. The scene as she sailed was a memorable one. Fully 100,000 spectators lined the landing stage and the river banks in the immediate vicinity and yelled themselves hoarse as the liner gathered headway down the river, and every steamer and riverside factory for miles along the Mersey joined in the chorus of goodbys [*sic*]. The din was deafening.

From the laying of her keel, to her launch, and those wonderful descriptions of her maiden voyage, scenes that were to be repeated on her arrival in New York, her story was just beginning. Here, through rare images and the meticulous research of the authors, the life history and memory of *Lusitania* is brought back to life, her full story and not just her tragic sinking. For me, *Lusitania* was one of – if not *the most* – beautiful ships ever created, from her unique vents that gave her upper decks a sleek and uncluttered profile to her graceful hull and towering funnels, she was a sight to behold, her interiors among the most luxurious, spacious and comfortable afloat. For me they rival those of the *Olympic*-class liners. These new and updated volumes are the Bible on the subject of the elegant and beautiful ocean liner RMS *Lusitania*, and bring her back into the spotlight that she truly deserves.

The Ghost of the Greyhound lives on.

Stuart Williamson, 2023

INTRODUCTION

New York City has long been known as the 'city that never sleeps'. However, on that night, things were certainly quieter than they would be a few hours later. The night was chilly and damp, with temperatures hovering around 48°F; rain had come through that morning, followed by fresh easterly winds; another round of rain was poised to move through the following morning.[1]

Over the preceding seven and a half years leading up to that rather cool Friday evening, a good friend to the city had made regular visits there. During that time, she had been so consistent and reliable, such a constant feature of life that her very presence must have seemed reassuring in a world that had fast become filled with uncertainties. Uncertainties like the so-called 'Great War' that had first embroiled Europe the previous summer, and then moved with lightning swiftness to become the first truly global armed conflict.

1 *The Sun*, 1 May 1915.

Lusitania rests quietly at her pier, 54, on the evening of Friday, 30 April 1915. (HFX Studios)

New York itself had not been thrust into the heart of the war. In fact, because America had not yet entered the conflict, there might have seemed times when what was happening in Europe could almost be forgotten – and yet, somehow, it could never be entirely forgotten, either. In a sense, things were not quite the same as they had been. There were concerns that the country would be drawn into the conflict – that last domino in the chain that had begun the previous year. That was a mere possibility, however. There were more tangible realities that gave evidence of the world situation: one was that many passenger liners now carried ammunition and war materials in their holds instead of just mundane items like lace, straw hats, soaps and perfumes, as they had before the war.

At the same time, because of the movement of war materials, European spies were also known to be operating all along the New York waterfront and wharves, monitoring shipping and clandestine cargoes on a variety of ships, and all reporting back to their superiors. And then there was talk of a newfangled weapon on the high seas, a contemptibly sneaky sort of craft that was growing increasingly concerning to many for the risks that they posed both to naval and merchant vessels, no matter what flag they were flying.

In short, tensions even in neutral, peaceful New York were running high that night, and with good reason. Yet the dark, four-funnelled silhouette that towered over the Lower West Side Manhattan waterfront, the funnels peeking up from behind the new Chelsea piers at the end of West 14th Street, was a reminder of better times. She was an old friend. A soft wisp of coal smoke drifted from her funnels as her boilers kept her electric generators humming away with a slight, energising tremor through her decks. A soft, warm glow emanated from every one of her illuminated ports, windows and deck lights. She certainly harkened back to a better time in the not-so-distant past.

In the seven and a half years that the British ocean liner *Lusitania* had been visiting New York, many things had changed. Yet it seemed to many that in spite of all the worries and concerns, fears and tensions, she would – she *must!* – remain a constant in the days ahead; that reliable companion that would never be found missing. Some genuinely believed that if North Atlantic storms and fog and icebergs and all that couldn't sink her, certainly nothing else could.

Yet on the night of 30 April 1915, some were concerned, too. The following morning, she was to sail back to a war-torn Europe, through a declared war zone. Although it was to be her 101st departure from New York to return to Europe, she had of late been running a gauntlet of sorts. Every other major liner on the North Atlantic had already left civilian service: White Star's 'ship magnificent' *Olympic*; the giant Hamburg-Amerika liners *Imperator* and *Vaterland*, the latter liner having been stuck at her pier across the North River for months now, plainly visible but totally immobilised by fear; even *Lusitania*'s own sister *Mauretania* had left the North Atlantic.

Lusitania would sail the next morning as both the largest and fastest liner then in service, with cargo holds laden with contraband that included munitions bound for the fronts in Europe. Yet some wondered and fretted: how long could she keep taking the risk and get away with it?

Only time would tell …

The Cunard Company's first steamship, *Britannia*, of 1840. (Authors' Collection)

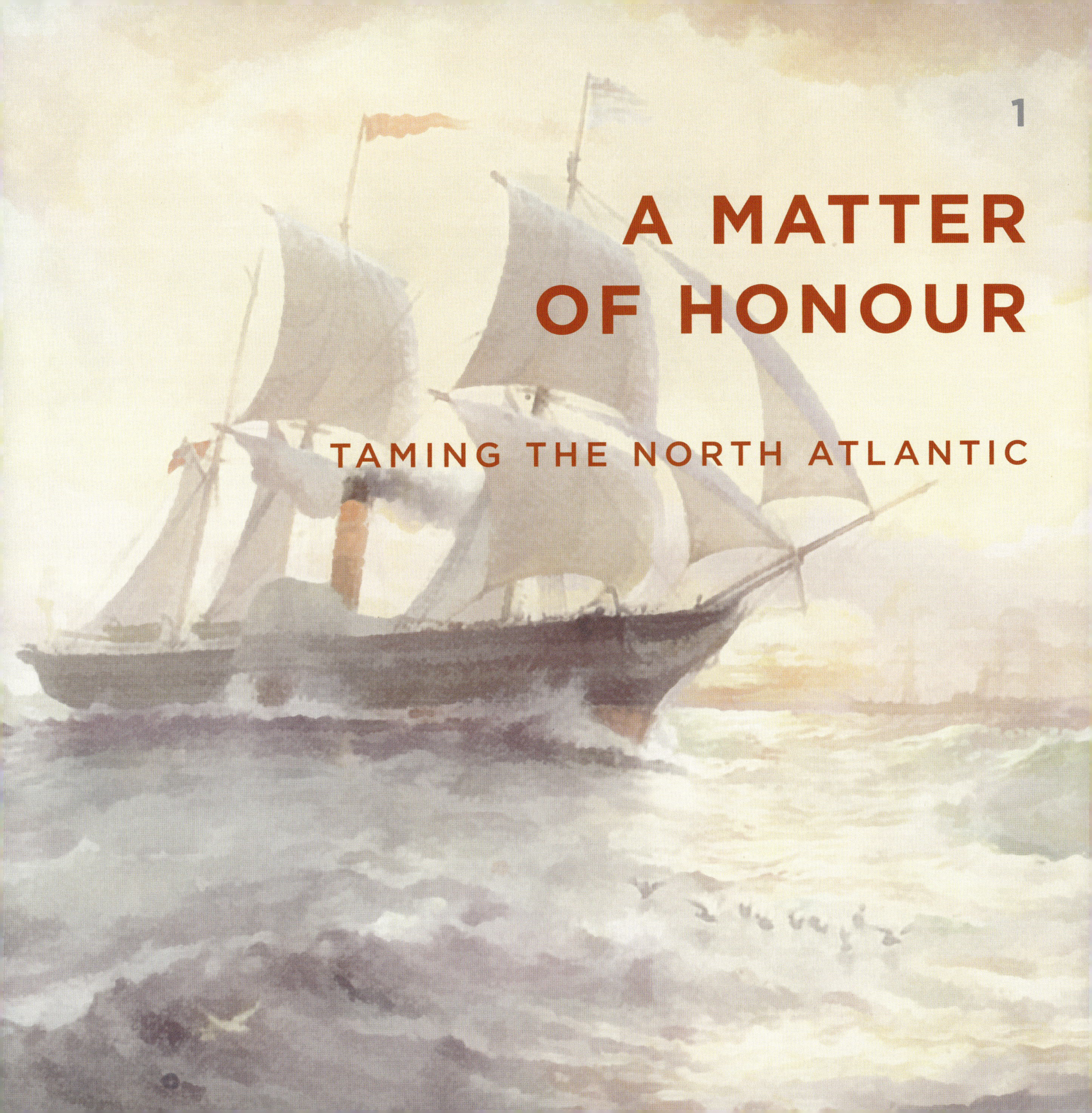

1

A MATTER OF HONOUR

TAMING THE NORTH ATLANTIC

Nearly nine years before that chilly night in 1915, in the late spring of 1906, New York had been both the same and a different place. It was always a busy city that seemed to be intent on progress; yet there had been more optimism in the air back then. The twentieth century had only just begun: a new age when things were supposed to get continually better. The times had felt momentous, and there were many newsworthy marvels of men to keep reporters busy.

The age of the skyscraper had truly come to life there as the century had begun. Names like 'Flatiron' and 'Singer' were watchwords of the day. Each one represented an enormous undertaking and a remarkable ongoing accomplishment by the human race. The dawn of the 'aeroplane' and the Marconi wireless radio, not to mention the rise of the horseless carriage, or 'auto-mobile', had kept the city enthralled. The music of the day was ragtime. Its king composer was Scott Joplin, an African American who would soon take up residence in the city. The popularity of his music proved just how far the nation had come in the preceding four decades, when a war had divided it primarily on the issues of race and slavery.

So on the day after the noteworthy event, it was not surprising that a story about it appeared in the pages of one of the world's most famous newspapers:

> WORLD'S LARGEST LINER THE *LUSITANIA*
> SUCCESSFULLY LAUNCHED AT CLYDEBANK
> GLASGOW, Scotland, June 7. – The new Cunard Line steamer *Lusitania*, the world's largest liner, was successfully launched at Clydebank to-day, and was named by Dowager Lady Inverclyde. Hundreds of visitors from all parts of the country, besides thousands of the local population, witnessed the ceremony.[1]

It would be quite some time, over a year, before *Lusitania*'s presence was actually felt in New York; the city had good reason to be excited about her, however, as the new Cunarder would eventually call the company's Lower Manhattan Pier 54 – then nothing more than a strip of concrete jutting into the North River with a small building sitting atop it – her home. As time passed and preparations leading up to her arrival continued, the New York press would really begin to whip up public excitement over the new liner.

1 *The New York Times*, 8 June 1906.

Hail, '*Lusitania*,' the first new leviathan twin to wed the ocean; fleet, strong and luxurious, a challenge to the world's mercantile marine. May she have a long career, maintaining prestige and profit to the nation and her Company.

Lusitania's Cunard Daily Bulletin, Vol. 1, No. 1, July 1907.

Naturally, the actual event had been far more exciting to those in Great Britain. The ship's launch at the Scottish shipyard had taken place at 12.30 p.m., and was followed promptly by a luncheon in the lavishly decorated moulding loft. There were over 500 VIP guests and dignitaries present. It was a monumental day for *Lusitania*'s builders and for her owners-to-be. Indeed, everyone in attendance had reason to feel proud over the accomplishment. The 'heartiest congratulations', of 'many notable shipbuilders and ship-owners from all parts of the Kingdom' were tendered. Sir Charles B. MacLaren, the Deputy Chairman of the builders, Messrs John Brown & Company, Limited, proposed a toast to the 'success of the *Lusitania* and Prosperity to the Cunard Company'.

William Watson, Chairman of the Cunard Company, rose to respond to Mr MacLaren's toast. Watson was new to the Cunard Chairmanship; the previous Chairman – and the driving force behind the idea for *Lusitania* and her sister – Lord Inverclyde, had passed away on 8 October 1905. Watson paused with a note of soberness to thank Lord Inverclyde's widow, Lady Mary Inverclyde, for having come down to the shipyard to ceremonially launch the ship. While it was 'one of the saddest of things' that he had not lived to see this triumph, Watson told Lady Inverclyde that the Cunard Board was grateful to her for honouring her husband's memory in agreeing to participate in the festivities, albeit 'at considerable cost to her feelings'.[2]

The British press was ecstatic with the latest maritime accomplishment of their country; even though *Lusitania* was merely an empty shell that day, a promise of what was to come, it looked as if they had every reason to be pleased. For a century, Britain had been the dominant maritime power in the world, both in naval terms and in the way of merchant trade, especially on the Atlantic. It was a matter of

2 *The Shipbuilder*, reprinted in *Ocean Liners of the Past*, Vol. 2, Mark Warren (Blue Riband Press, New York, 1997), pp.15–16. [Henceforth cited as *Shipbuilder*.]

Far left: George Arbuthnot, Second Baron of Inverclyde and grandson of Sir George Burns, a co-founder of the Cunard Company. It was under Inverclyde's guidance that plans to build *Lusitania* and *Mauretania* were conceived and construction commenced, but he passed away before *Lusitania*'s launch. (Authors' Collection)

Left: Sir William Watson, Chairman of Cunard when *Lusitania* was launched. (Authors' Collection)

necessity since Britain, an island nation, depended on trade and naval power. A British statesman reportedly once said that only through ships did England 'exist as a nation', adding: 'England owns the Atlantic – the Atlantic is a British pond.' This was no exaggeration.

There was a reason for that self-assured smugness, at least on the face of things. Particularly through the latter half of the nineteenth century, Britain had stared down some potentially worrisome competition from other nations. During those years, the ocean liners crossing the North Atlantic had all seen remarkable technological advancements. They had gradually gone from small, weak vessels powered by sail, to frightfully underpowered steamships still ready to hoist their sails if their engines failed them or they ran out of fuel. Then the ships had grown larger, more comfortable, more reliable, and faster. With each advance in seagoing technology, the transatlantic passage across 'the pond' had slowly turned from a weeks-long ordeal with the possibility of being drowned into a regular routine filled with unprecedented comfort and ever-increasing safety. Potential passengers were advised to take a sea voyage 'as a rest and air cure ... Taking it all in all, sea trips [were] very strongly recommended as important hygienic factors,' for the 'invigorating effect of the ocean climate'. One American book of the era put the popularity of taking a 'foreign tour' to Europe this way:

> But a few years ago a tour Around the World was beyond the reach, and almost beyond the ambition, of all save the very wealthy. Today, however, up-to-date machinery has almost annihilated time and distance, so that it is now one of the easiest, and certainly the most charming way of acquiring practical knowledge of other countries and other peoples under the most delightful conditions.
>
> ... The foreign tour has now become a yearly function of the large contingent of American pleasure-seekers, and the volume of travel is steadily increasing year by year. Within a very few years pleasure travel Around the World has more than quadrupled ... Every Spring a huge exodus sets in, and then everybody seems to be going abroad. Friends and acquaintances plan ahead

> and make appointments from one year to another for reunions, etc., in London, Paris or other cities.[3]

Because of the technological advances in maritime construction, passengers were informed that they no longer had to worry about the dangers of an ocean passage:

> After fighting the sea and its terrors for thousands of years, man has at last succeeded in conquering the sea, this wildest and most unruly of Nature's children. Against the modern iron or steel ship, which is equipped with every measure of protection that science and engineering can devise, the sea is almost powerless. Smaller vessels and sailing craft still feel its fury occasionally, it is true, but the enormous ships of the present day forge their way through the mighty ocean at high speeds.[4]

Competition on the North Atlantic had really begun to develop along national lines during the last decades of the nineteenth century and the first decade of the twentieth century, and the whole thing had become a matter both of honour and of national pride. Although the French, Germans and Americans all had shipping interests and respectable steamship companies running on the Atlantic, it was really the British companies that had set the benchmarks for this trade. Halifax businessman Samuel Cunard had inaugurated the Cunard Company in 1840 to capture a Royal Mail subsidy.

That postal subsidy from the British Government gave their shipping the edge in the Atlantic, which in the previous generation had been dominated by the American sail-powered packet trade. Despite the Americans being the first to find commercial success in steamers, with Robert Fulton's *North River Steamboat*, and being the first to deploy steam power on the transatlantic route with the ship *Savannah*, the rest of American shipping was slow to adopt the new technologies. With the Royal Mail backing, Cunard was an unstoppable force focused on recapturing British supremacy.

The US shipping industry had responded with two conflicting approaches. America's first approach relied less on subsidy and more on private industry, with shipping and railroad mogul Cornelius Vanderbilt creating the Vanderbilt Line. The second response came from Edward Collins, son of the founder of one of the foremost American packet companies; he had spent a decade petitioning the US Government to create a postal subsidy of their own to compete with Britain, which ultimately enabled him to create the overly lavish and tragically short-lived Collins Line. The Collins Line's heavy reliance on a reluctant government subsidy was their ultimate downfall, only exacerbated by the disastrous loss of two of their ships, *Arctic* and *Pacific*.

While the greyhounds battled each other on the North Atlantic's blue waters, on the shore the greycoats battled for secession against the bluecoats. The American Civil War tore the country apart, significantly crippling American shipping. Even after the war, America's attention diverted from outward expansion to inward development.

Cunard, however, still had not been free from competition. In early 1868, a dying packet ship company – one that actually dated back nearly as far as Cunard – was revived by a British steamship maven, Thomas Henry Ismay. This company was the White Star Line, which began directing its attention to developing advanced transatlantic ships alongside their partner shipyard, Harland & Wolff. Cunard and White Star, despite being rivals, became the premier shipping companies in the world.

One of the worst disasters in the nineteenth-century North Atlantic befell the White Star Line when their steamer *Atlantic* crashed onto the rocks of Nova Scotia in 1873. *Atlantic* was the second ship of Ismay's White Star Line and was a total loss, resulting in the deaths of around 550 people. While the Collins Line had been severely hindered by their shipwrecks, the White Star Line managed to press onward and recover both in reputation and finances.

When White Star's *Teutonic* entered service, she was the pride of Great Britain's mercantile fleet. On 3 August 1889, while the new liner was participating in the Spithead Naval Review, she was inspected by the Prince of Wales and Kaiser Wilhelm II. Kaiser Wilhelm was Queen Victoria's grandson through her daughter Victoria, Empress Frederick of Germany. Born in 1859, Wilhelm was only 30 when he boarded *Teutonic*. Young and ambitious, Wilhelm was dearly devoted to his grandmother, but was also eager to see his own nation become a rival to Great Britain – perhaps even to usurp her position of dominance on the world scene. While touring *Teutonic*, he was apparently heard to remark: 'We must have some of these.'

---

3 *Outward Bound: Twentieth Century Ocean Travel to Fascinating Far-away lands*, by John H. Gould (Outward Bound Company, 1905/1913) pp. 1-2. (Henceforth cited as *Outward Bound*)

4 *The Scientific American Handbook of Travel*, (Munn & Co., 1910), pp.2–3.

North German Lloyd's *Kaiser Wilhelm der Grosse* of 1897 was the first in a series of German liners that dealt a blow to British maritime prestige on the North Atlantic. (Authors' Collection)

And in short order, they did 'have some' of them. Germany had long had two main merchant lines, Hamburg-Amerika and Norddeutscher Lloyd (North German Lloyd); although they had been successful companies, they had not carried much in the way of prestige like the White Star or Cunard ships. Yet in 1897, North German Lloyd fulfilled the Kaiser's dreams when they introduced *Kaiser Wilhelm der Grosse*, named in honour of the Kaiser's own grandfather. *Kaiser Wilhelm* was the longest and largest liner then afloat, and also proved to be the fastest, quickly taking the Blue Riband – the much-coveted yet non-corporeal prize for the fastest ship on the Atlantic – away from British liners. With this development, British maritime confidence suffered the first in an unfortunate series of body blows: following quickly in *Kaiser Wilhelm*'s wake was a string of German liners that were the largest, fastest and most prestigious ships of the day.

Then came another setback to British prestige: in 1902, American financier J.P. Morgan bought out the White Star Line, solidly placing one of the two great British lines under American ownership. Morgan also tried to buy Cunard, but Cunard Chairman Lord Inverclyde was firmly opposed to the takeover. His company's position was unenviable, however; Morgan's new combine, the International Mercantile Marine, now had the prestigious White Star Line at the fore. They were planning new and startlingly luxurious ocean liners for the Atlantic service. At the same time, the German lines held the Blue Riband, sitting comfortably as genuine trend-setters for speed and over-the-top luxury. Then a rate war heated up, with competitors slashing prices at a speed that Cunard could barely keep up with.

Cunard's front-running liners just before the turn of the century were *Campania* and *Lucania*, both of 1893 and quite modest in comparison with the German and White Star ships. Cunard could not maintain their prestigious position on the Atlantic without putting new superliners into service, but did not have the resources. Their new *Ivernia* and *Saxonia* were the largest Cunarders ever built when they entered service in 1899 and 1900, respectively, yet they still could not compete. Much more had been needed.

In what was now a matter of honour, *Lusitania* was to send a clear message to Germany and the aggressive Kaiser Wilhelm II. In creating this new vessel, Britain had given birth to a new era in ocean-going technology. Oddly, *Lusitania*'s launch in a sense brought the world one step closer to an uncertain future, a future that would include a full-blown world war. On that Scottish late-spring day of 7 June 1906, however, the word '*Lusitania*' meant only one thing: progress.

How had Cunard gone from their tenuous position in 1902 to launching the largest ship the world had ever seen, with another nearing launch, in just four years? Virtually everything had gone exactly the way it had needed to go in order to make the dream a reality.

Lusitania fitting out at the John Brown shipyard on the River Clyde, as seen from the opposite shore. (J&C McCutcheon Collection)

2

ORIGIN STORY

BUILDING *LUSITANIA*

Lord Inverclyde had apprenticed in various departments within Cunard, and had a broad knowledge of how to run a shipping line; he was well known as a thinking sort of person who could come up with ideas to face critical situations. To construct ships that would respond to the German and White Star threat, he needed a financial backer. The deepest pockets available were controlled by Parliament, so in March 1902, Inverclyde approached the British Government for assistance. Appealing to feelings of bruised patriotic prestige, Inverclyde proposed that the British Government provide Cunard with a loan to give them the fiscal fluidity needed to construct the largest and fastest vessels in the world. These vessels would, in turn, recapture the mercantile pride from Germany.

No earthly government moves quickly, and the proposal took time. In July, the stewardship of the Government was changing, as Prime Minister Salisbury had been succeeded by his nephew, Arthur Balfour. Balfour's younger brother Gerald was then president of the British Board of Trade, and this familial relationship seemed to help pave the way for an agreement with Cunard. On 30 September, Prime Minister Balfour announced the agreement, and Gerald Balfour gave a public speech in Sheffield where he discussed the threat from German shipping lines, pointing out that Cunard was incapable of funding a response without assistance. He then announced a tentative agreement with Lord Inverclyde, which he expected would soon be approved by Cunard's shareholders. Despite interruptions of applause from his enthusiastic audience, he was able to convey that not only had Cunard pledged to remain an all-British company crewed by British officers, but that they would also undertake to build 'two vessels of 24 to 25 knots' which would be available to the Admiralty and the Government if their services were needed. He also announced an annual £150,000 subsidy for the operation of the two ships.

Arthur Balfour, British Prime Minister from 1902 to 1905. (Authors' Collection)

On 30 July 1903, the final agreement was signed. It stipulated that the Government would supply a sum not to exceed £2.6 million in the form of a twenty-year loan, at an interest rate of 2.75 per cent – less than half of the going rate – in order to fund the construction of the two vessels. Additionally, they would pay the aforementioned £150,000 annual subsidy to Cunard for the ships' operation and upkeep. Under this agreement, Cunard was to remain a British-only operation, with 'no foreigner' being qualified to serve as a director or principal officer of the company. Furthermore, no shares of the company were to be held by, 'or in trust for, or be in any way under the control of any foreigner or foreign corporation, or any corporation under foreign control'. The agreement also included a mail contract.

On the other side of the equation, the Admiralty was to have complete approval over the ships' design. Each was to maintain a minimum average speed of 24½ knots in moderate weather. If the speed of the new ships fell below that mark, but did not fall below 23½ knots, then a deduction from the annual £150,000 payment would be made. A speed of less than 23½ knots could mean further penalties, or even the complete rejection of the ships. The new liners were to be liable for charter by the Government in time of war, and were to be designed for conversion into high-speed, albeit unarmoured, auxiliary cruisers with formidable armament, should the need arise. Additionally, the charter of all Cunard ships by the Government was also assured. Once Cunard had signed on the proverbial dotted line, everything was ready for Cunard to proceed with the designs of and orders for two new sister vessels, the likes of which the world had never seen.

Although much work had already been done on the project even before the agreement was finalised, still more lay ahead before the first ship's maiden voyage could commence. Time was of the essence for Cunard: the more time passed, the greater their competitors' lead would be. Instead of waiting until the British Government had made a final agreement with them to fund the new ships, Cunard had long since begun working on initial plans for the liners.

As far back as November 1899, Cunard had requested a quote, or tender, from the John Brown shipyard at Scotland's Clydebank for a single ship of 700ft in length by 71ft wide. By 1901, Cunard had seen the need to build a pair of ships to keep up with the growing competition; Cunard's General Superintendent James Bain and Naval Architect Leonard Peskett, had drawn up a rough concept for 700ft liners capable of speeds up to 24 knots. Things had really gone nowhere, however, at that point.

In February 1902, Cunard had approached three prominent shipbuilders – John Brown & Company of Clydebank in Scotland; the Fairfield Shipbuilding & Engineering Company of Govan, Scotland, also located on the Clyde; and Vickers Sons & Maxim, Ltd, of Barrow-in-Furness – with the idea of getting quotes for a pair of ships. In October, they finally received a full report on the companies' submissions.[1] Based on these reports, which indicated the project was feasible, Cunard drew up new draft specifications and general arrangement plans, and then submitted these to the three firms. A fourth shipbuilder was also included, C.S. Swan & Hunter at Wallsend, after that firm's Chairman, George Hunter, convinced Cunard to allow his company a chance at the contract.

This English shipbuilding firm had not yet broken into the list of top-notch firms of the day; however, they had already built for Cunard the intermediate steamer *Ultonia* of 1899, of some 8,845 gross registered tons (grt), and also *Ivernia* of 1900 and 13,799grt. *Ivernia* had been paired up with the John Brown-built *Saxonia*; the order for a third ship of the class, *Carpathia*, had been placed with Swan & Hunter, so the shipbuilder seemed to be in an increasingly favourable position with Cunard.

When the Government had given their conditional agreement to Cunard in 1902, the line's directors found that the originally proposed specifications for the steamers were no longer adequate, and went back to the drawing board. By early 1903, the working idea was to build two ships of 750ft between perpendiculars (bp)[2] by 76ft wide; additionally a speed of 25 knots, instead of 24, was now the goal. For the sake of practicality, the construction of each of the two new ships was to be farmed out to a different yard; this would allow for simultaneous construction.

On 18 December 1902, Cunard's Board of Directors reviewed the four proposals. The most attractive submissions were by Swan & Hunter, for a ship of 760ft bp by 80ft, and Vickers, which proposed a ship of the same length, but with a slightly slimmer beam of 78ft. Each ship would be powered by three sets of tandem, quadruple-expansion, five-cylinder reciprocating engines outputting over 60,000hp. Models of these forms were built and tested, and the Swan, Hunter design was found to achieve the same speed as the Vickers design, but required 7 per cent less horsepower. John Brown, meanwhile, had proposed a ship of 725ft bp by 80ft wide, but it did not model test very well. A series of model tests were then carried out under the supervision of Sir Philip Watts, the Director of Naval Construction to the Admiralty. The final result of the tank tests, which concluded in April 1904, was for ships of 760ft bp by 87½ft in width with a working draught of 33½ft.[3] This specification forced Vickers out of the running, ostensibly because they did not have facilities to build a ship of that width.

However, nearly a year earlier – at an 8 May 1903 meeting with representatives of the prospective shipbuilding firms – Cunard had already shown a preference to select John Brown and Swan & Hunter. As long as the two companies could 'collaborate with the view to producing duplicate ships', they were considered the clear choice.[4] Before May 1903 was out, and in the hopes of securing the prestigious contract, Swan & Hunter had merged with the nearby shipyard of Wigham Richardson in order to form the Swan, Hunter & Wigham Richardson Company.

Meanwhile, the Clyde River, which both John Brown and Fairfield yards fronted, was inadequate to handle vessels of the size being considered. In order to seriously pursue the contracts, the companies informed the Clyde Navigation Trust – a public statutory body responsible for maintaining and improving that river – that a contract for one of these

1 University of Liverpool Archives, D42/B8/1; *Ships for a Nation*, p.109.

2 A ship's length between perpendiculars is a measurement between the forward and aft perpendiculars. The forward perpendicular is → a vertical line that intersects the forward side of the stem at the waterline; the after is a similar vertical line that intersects the after edge of the stern post at the waterline. The actual overall length of a ship would include additional portions of her structure that would extend beyond each of these points.

3 *Shipbuilder*, November 1907, p.5.

4 *Ships for a Nation: John Brown & Company, Clydebank*, Ian Johnston (Argyll Publishing, 2008), p.111. [Henceforth cited as *Ships for a Nation*.]

liners could not be accepted 'unless there was an absolute certainty that the channel leading from the yards concerned to the sea was deepened and widened to a certain extent'. The Trustees 'at once gave their assurance that there would be sufficient water by the launching date'.[5]

The pieces were beginning to fall into place. On 30 April 1904, Cunard's directors met with representatives from Swan, Hunter and John Brown to inform them that they should 'regard the order as having been placed' with them. Both shipyards immediately began work on their respective projects. Incredibly, as each yard began work on their respective build, there were no signed contracts for the orders; some of the vessels' specifications were either still undecided or in need of Admiralty approval. The final contracts were signed in London on 18 May 1905, stipulating delivery of both ships within thirty months from the date of signing, or by 18 November 1907.[6] Both liners were to be built on a cost-plus basis rather than for a fixed price, with Cunard paying for 1) the cost of materials, parts and labour; 2) 15 per cent profit on those expenses for the yard; 3) the cost of delivery to Liverpool, drydocking, coal for the trials, etc.; and 4) 5 per cent profit on the preceding three totals. Each shipyard was to submit a monthly bill to Cunard, which would in turn be submitted for a monthly draw from the Government loan.

Frighteningly, there should already have been indications that trouble lay ahead. No shipyard had previously built ships of this ilk, and the newly amalgamated Swan, Hunter was particularly inexperienced at undertaking such a large-scale project. Precisely estimating costs for the ships' engines was going to be difficult, since the scale of application, desired horsepower and speeds being aimed for were all unprecedented. Lastly, and perhaps most alarming to anyone who was thinking ahead at all, on 23 April 1904, Swan, Hunter had submitted an estimated cost for their build of £1,301 million, already over half the loan total, *excluding* the costs of interior decoration. The situation would not improve from this point, and in the end both liners would experience massive cost overruns as they approached completion.

All of that was still over the horizon, however, when work began. Before the first keel plate of John Brown's ship could be laid, an appropriate slip to accommodate her unprecedented size needed to be built. Since the Clyde River

5 *Glasgow Today*, 1909, p.23.

6 *Ships for a Nation*, p.111; *American Machinist*, Vol. 31, Part 1 (5 March 1908), p.351.

DECIDING ON TURBINES: THE *CARONIA* AND *CARMANIA* LEGEND

It is often said that the final decision to use turbines on *Lusitania* and *Mauretania* came only after *Carmania*'s turbine powerplant proved superior to the reciprocating engines used on her nearly identical sister *Caronia*. This is untrue. A simple timeline shows the problem with this concept.

It has often been said that the reason why work was focused on the bow and midship regions of *Lusitania* during construction had to do with waiting for the final decision on whether to use reciprocating engines or turbines; however the decision to use turbines had been made six months before *Lusitania*'s keel was laid. Quite simply, *Carmania*'s performance in December 1905 had no bearing on the decision to use turbines on *Lusitania* and *Mauretania*, made in March 1904, over a year and a half earlier.

1901

Late 1901: Cunard approaches Charles Parsons over the possible use of turbines. Parsons recommends that Cunard use turbine machinery in their new ships.

1902

1902–early 1903: Due to a lack of experience in using turbines, Cunard tacks back towards using reciprocating engines, since they were tried and tested technology.

1903

20 August 1903: A special Turbine Committee is formed to consider which type of propulsion would drive the new sister ships; it includes Charles Parsons as well as representatives from Cunard, the Admiralty and both John Brown and Swan & Hunter.

fronting the shipyard was only slightly over 600ft in width, it was far too narrow to launch *Lusitania*'s nearly 800ft hull into. However, the frontage of the yard also came into close proximity with the confluence of the Clyde and Cart Rivers. Thus, the new slipway – replacing two previous slips – was built on an angle of 40 degrees to the line of the river, to allow a comfortable launch run of 1,200ft.

With these preparations complete, it was time for the real work to begin. The first keel plate for John Brown Hull No. 367 was laid down on Wednesday, 17 August 1904 – a ceremony that Cunard's Chairman, Lord Inverclyde, reportedly participated in.[7] From that time, work progressed quickly, and soon nearly the entire keel had been laid. Initial work was focused on the forward and midship regions of the vessel, as certain minor details of her final configuration – such as the positioning of bulkheads and some of the machinery seatings – had not yet been finalised.

As construction commenced, the ship – and her future running mate taking shape at the Swan, Hunter shipyard – was yet without a name. It was not until 15 February 1906, at a meeting of Cunard's Board of Directors, that names were formally selected. Among the contending pairs of names were *Britannia* and *Hesperia*, *Britannia* and *Hibernia*, and *Albania* and *Morovia*. In the end, the names *Lusitania* and *Mauritania* were selected. John Brown Hull No. 367 was to become *Lusitania*, while the ship being built by Swan, Hunter would be named *Mauritania*. The names were selected in honour of the names of former Roman provinces located in Spain and Africa, respectively, and bore Cunard's traditional *-ia* suffix. The choices were soon made public, but Cunard quickly realised that the spelling of the Roman province whose name the Tyne ship bore was not *Mauritania*, with two 'i's. Instead, the correct spelling was *Mauretania*. They made the necessary alteration, thus confusing generations of maritime enthusiasts more familiar with the spelling of the modern nation Mauritania.

While Cunard was deciding on names, the John Brown yards were alive with activity; pride in the project was running very high. This yard, these workers, were participating in the effort to build the world's largest and fastest ship – right there in Scotland. Naturally, *Lusitania* was the pride of the British Empire; however, it was only natural that Scottish people felt a particularly strong attachment to her.

7 University of Glasgow Archives; *Ships for a Nation*, p.337. For many years, *Lusitania*'s keel-laying date has been incorrectly recorded as 16 June 1904.

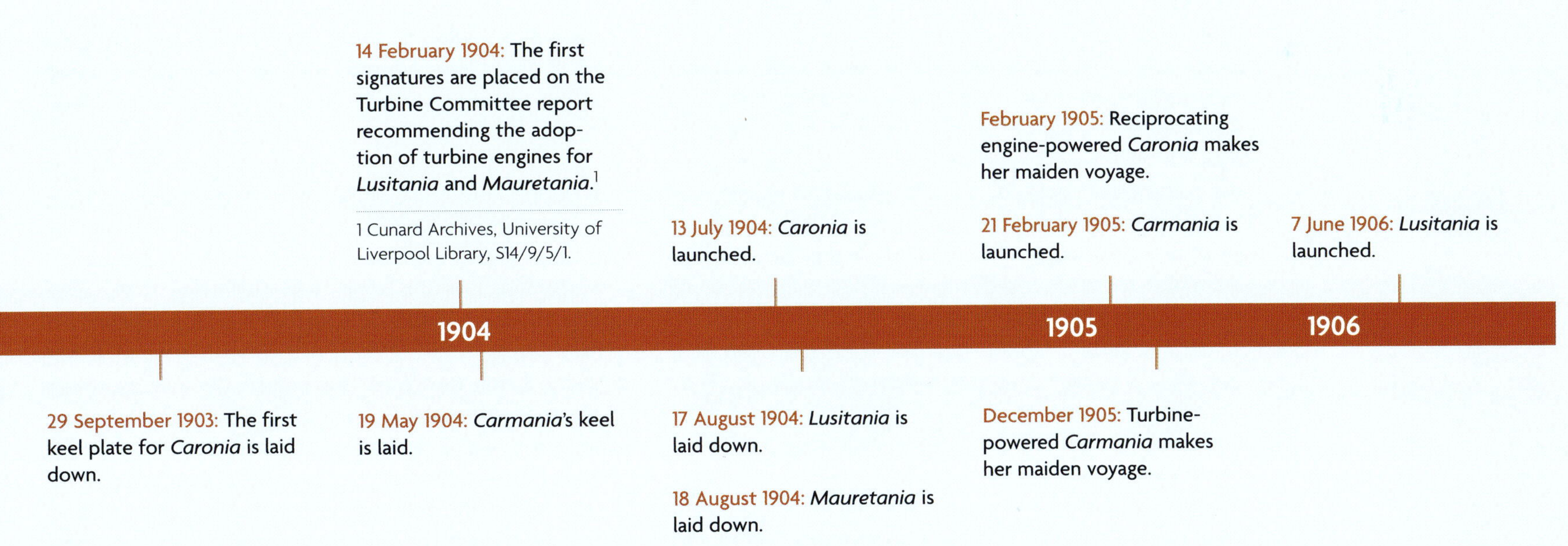

The stern form of *Lusitania* taking shape. The ship's stern post, the mount for the rudder and the after propeller bossings are all clearly visible. (Authors' Collection)

The way they felt about *Lusitania* is perhaps best summed up by a statement that one of her Scottish engineers made just as her Boat Deck was about to slip under the waves in 1915: 'Och,' he reportedly exclaimed, 'but they can't sink a *Clyde-built ship!*'

At the English yards of Swan, Hunter, work on *Mauretania* was ongoing. Although her keel was laid down only a day after *Lusitania*'s, she was progressing more slowly than her Scottish-built sister. Whereas *Mauretania*'s construction was carried out underneath a large steel and glass enclosure, protecting workers from the often inclement English weather, at John Brown work continued in the open air. From a distance, it actually looked more as if *Lusitania* was simply sprouting from the muddy riverbank.

What is certain is that Cunard was hoping to learn from operating a turbine powerplant through *Carmania*. They gained crucial experience that, in the end, improved the final success of *Lusitania* once she entered service in September 1907.

Lusitania's construction meant that the John Brown yard would be bustling with work for a full three years, regardless of other simultaneous projects. However, there was also labour unrest at the time; there were still several occasions when work on the ship was stopped due to strikes. One of these, in late 1906 by the boiler production crews and steel workers, lasted seven weeks. During this strike, work was carried out by their apprentices until it ended and work continued.

Despite this, work on *Lusitania* proceeded at a very good pace. By the late spring of 1906, her hull towered above the shipyard and the Clyde River. Since there was no local drydock large enough to accommodate *Lusitania*, the ship would have to sail down to Liverpool, England, for her first drydocking; thus, all four propellers were mounted prior to the launch. To protect the delicate blades during the procedure, they were bound carefully in temporary casings.

Preparations for the launch were extensive. At the bow, a strong cradle and fore poppet had been constructed and affixed firmly to the shell plating to give the ship added stability during the critical moments when her stern was afloat and her narrow bow was balanced on the ways. They were fitted with tripping lines and bolts so that once the ship was afloat the entire structure could be removed easily. It was against this fore cradle that the hydraulic cylinders would be used if the ship did not begin to move under her own weight when the time arrived. As the time for the event neared, the keel and bilge blocks were progressively

A comparison of two nearly identical photos of *Lusitania* before her launch. The upper photo shows scaffolding along the upper structure as workers complete preparations for the launch. The propellers are being installed. In the lower photo, the ship stands nearly ready for launch. (Authors' Collection)

Above: A comparison of two photos showing the stern on the slipway. In the left photo, work is still under way around the propellers and rudder, and scaffolding still hangs over the sides of the ship. In the right photo, the ship is ready for launch. (Authors' Collection)

Left: The presence of a woman in the shipyard indicates that this was likely the morning of the launch, 7 June 1906. *Lusitania* towers over everything and everyone present. (Authors' Collection)

removed until the ship was supported on the ways by the cradle alone.

The standing ways were positioned between the centre keel blocks and the outer bilge blocks; they were lubricated with a ⅜in-thick layer of Russian tallow and soft soap, and the 680ft sliding ways were coated similarly. Six electrically operated triggers, three on each side, would start the ship on its journey to the water.

Getting the hull moving was one thing; stopping its incredible momentum was another. Without some sort of stopping mechanism, she would simply careen across the full width of the waterway and end up on the opposite bank of the river. Over a thousand tons of drag chain were set up to act as a brake. The chains were grouped in carefully arranged piles, attached with wire cables to the hull's shell. They were spaced at roughly 10ft intervals and would come into play sequentially.

On the day of the launch, Thursday, 7 June 1906, the John Brown shipyards were buzzing with activity. The weather was sunny, all in all a beautiful late-spring Scottish day. While workmen carried out the final launch preparations, careful inspections were made all around the ship to ensure

The bottle of champagne has been sent on course for the ship's prow. Within seconds, the liner would begin sliding down the ways. (National Archives & Records Administration, Authors' Collection)

that everything was as ready as it could be. Well-dressed, specially invited spectators began arriving, some 600 all told. Some of these were allowed to venture down on the bank of the river, practically underneath the hull, to get a good look at the liner as it towered over them. Members of the press began arriving in droves, and thousands of spectators lined the banks of the Clyde River along both sides, as well as on the decks of waterborne craft that, as the time approached for launch, stood off at a respectful distance from the 16,000-ton hull that would soon be entering the river.

Anticipation in the yard mounted as the predetermined hour approached, and everyone present took their places; all eyes were fixed either on the unbelievably large hull, or on Lady Inverclyde, who was standing on the launching platform near the bow. At the set time, 12.30 p.m., Lady Inverclyde pushed the specially rigged button that tripped the triggers and smashed the bottle of wine against the prow, christening the unfinished ship '*Lusitania*'. For the following twenty-two seconds, there was a tremendous creaking and groaning noise as the hull's weight shifted upon the ways. In the end, simple physics told the enormous structure which way it had to go: the ways were lubricated and the hull could not defy gravity, and so the only option was to follow the declining ways into the water. During this twenty-two-second time of decision, the hull moved only 1ft.

Then, during the subsequent sixty-four seconds of the launch, the hull began to move in earnest. The band present for the occasion struck up 'Rule Britannia' on cue, and as she slid towards the water the Union Jack flapped proudly at the prow. It was, as one journalist later pointed out, an event of 'national importance', not only because this was the greatest maritime achievement in history, but also because this ship would hopefully best Germany.

It must have been a spectacular sensory experience; the bright sunshine and blue skies showed the scene in brilliant colour. The white, black and dark red of the hull gleamed in all its newness as it ran towards and finally parted the waters of the Clyde. The sound of the ship moving over the ways was thunderous, with the hull groaning while the

Lusitania taking to the water for the first time. Piles of drag chains once used aboard Brunel's giant, *Great Eastern*, are ready to check the hull's gathering momentum. The first group would come into play as soon as the ship had fully travelled the ways, while the heaviest pile would come into action last. (Authors' Collection)

ways creaked under their respective stresses; next came the sound of the giant chains jangling as their connecting stays were brought taut, one after another, and the piles began to drag along the riverbank. Then, in a moment, the launch had ended. The hull's movement stopped, the noise of the chains ceased, and only the cheers from spectators and the playing of the band remained. In the middle of the Clyde, the hull bobbed serenely in the water; without any fittings or machinery aboard, she rode unusually high, with the tops of her propellers yet exposed.

In the primary sixty-four seconds of the launch, the ship had reached a speed of 12.2ft per second. The last pile of drag chains did not even see use before the hull stopped in mid-river. Six tugs quickly took positions around the giant form, attaching lines to the hull so that she could be warped into her fitting-out basin for the final stages of her construction. The pins holding the drag chains to the hull were released, and *Lusitania*'s hull was brought safely to the fitting-out basin, completing her first voyage successfully.

Although everyone celebrated and newspaper reporters throughout England thrilled readers with tales of the event, the truth was that the launch was only the beginning of the second half of *Lusitania*'s construction. Work began afresh on the ship the next day, turning her into a finished, seaworthy ocean liner.

Fitting out *Lusitania* was an enormous task since she was then a mere shell. There were no engines, boilers, machinery or equipment. Each deck was only a steel floor and ceiling, with columns supporting the decks overhead; there was no internal subdivision other than the watertight bulkheads found below. Outside, the superstructure was only complete up to the floor of the Promenade Deck, while forward, a mere fraction of the Promenade Deck's enclosure and the flooring of the Boat Deck was in place. There was nothing above these points – no funnels, no masts ... nothing. All of these items, as well as the ship's lavish appointments and furnishings, would be added during the fitting out.

The first step was to take *Lusitania*'s machinery from shore, hoist it aboard the liner and put it into place. All of the boilers, uptakes, turbines, condensers, etc., had been produced ashore. With the ship now ready to accept the

Lusitania's hull, safely launched, rides high in the water of the Clyde. Her prow rests 110ft from the end of the ways. She would soon be towed to the fitting-out basin. (HFX Studios)

A portion of one of *Lusitania*'s turbine casings arriving at the shipyard, where workers pose proudly beside it. This view gives some sense of scale for how big the liner's turbines really were. (Richard Smye Collection)

machinery, each piece, in turn, was placed on carts that ran on railway-style tracks through the shops and out onto the wharf beside where *Lusitania*'s hull was berthed. Once alongside the ship, a large crane would hoist the item from the cart into the air. This crane had been built by Sir William Arrol & Co., Ltd. It stood on the east side of the fitting-out basin, and had a lift capacity of 150 tons (Imperial).[8] Once the item was in the air, a delicate choreography followed: all along the liner's decks, large openings gave access to the lowest regions of the ship. Eventually, these would be sealed, used for funnel uptakes and ventilation spaces. With them still open, however, the crane operators could carefully lower the large machinery into the bowels of *Lusitania*'s hull, whereupon it was guided cautiously into its final mounts and secured to the ship. Eventually, all of the boilers, turbines, condensers and machinery found their way aboard. Following this, the large funnel uptakes and boiler flues were hoisted into place and attached, sealing the machinery in place.

8 *Engineering* [reprint], p.61.

RECREATING A GREYHOUND

Most of the artistic pieces that appear in this two-volume set come from a highly detailed three-dimensional computer model of *Lusitania* built for the HFX Studios' virtual museum experience *Lusitania: The Greyhound's Wake*. Our project to recreate *Lusitania* began in June 2019, headed up by HFX Studios founder Tom Lynskey and Lead Animator and Modeller Levi Rourke. In short order, maritime historians and authors J. Kent Layton, Tad Fitch and Michael Poirier were brought in as advisors. Although he joined the project much later on, Digital Artist and Modeller Alex Moeller brought with him an incredible skill set and attention to detail.

Once the project began, the members pooled their own extensive available historical data and references on the ship. 'Mike Poirier was constantly turning up these really rare articles and images of the ship and sending them over to the modelling team,' Layton recalled. 'It was amazing to open my inbox every day and see what he had turned up since the previous night.' Tad Fitch brought an extensive collection of liner photographs, letters, artefacts and ephemera; he had also spent decades researching *Titanic* and other liners of the era, including the submarines and events of the First Battle of the Atlantic during the First World War.

The team was not content with their own data. They quickly contacted a number of maritime experts, collectors and artists from around the world, and were generously granted access to consult an unprecedented wealth of data. At different points during the project, they were assisted by the likes of Mark Chirnside, Stuart Williamson, Bill Sauder, Ken Marschall and Eric Sauder, each of whom had unique insights, perspective or research material to add. All this information was compiled to see how *Lusitania* was riveted together and how she had changed throughout her career.

Tom Lynskey said, 'We had recently come off of working on the *Olympic*-class ships, which are simpler in both architecture and decoration. We were in uncharted territory with *Lusitania*, however; no one had ever sat down and scrutinised her like this, and we quickly realised how complex a vessel she really was.' Although both Lynskey and Rourke had previously been involved in bringing the White Star liners *Titanic* and *Britannic* back to life in a similar fashion, they quickly began to learn that *Lusitania* would present its own unique challenges. 'Having worked on two larger ships in the previous projects, *Lusitania* proved to be far more complex despite being slightly smaller. There was never a space that *didn't* cause much frustration with the subtle sheer and camber acting on every wall and floor,' Rourke pointed out.

Lusitania was designed not only with sheer, that graceful vertical sweep fore and aft, but also with a very deep camber. This was a side-to-side crown of the deck, and *Lusitania*'s was far greater than the White Star liners they had previously worked on. While working on their previous projects, *Titanic* and *Britannic*, there was not even an attempt to model camber. The combination of the two curves was shockingly complicated to model on *Lusitania*'s already complex shape, and every room had to be built on the combined curves. 'It was well worth the extra work,' Layton points out. 'Although the sheer and camber is very subtle, you can really tell it's there, and it really brings the ship to life.' Tom emphasises that there's hardly a straight line anywhere aboard.

Once the basic shapes of the ship and her rooms were set in place, the team had to turn them into the grand spaces that they originally were. Digital artist Alex Moeller joined the project when it was already in progress. 'When Alex joined,' Layton recalled, 'some of the early artists who had done preliminary work and temp pieces had left to pursue their own projects. Alex really rolled up his sleeves, bringing an unparalleled level of skill and artistry to the project; he began focusing on the nitty-gritty where the others had left off. When Alex got to work, the details really started to pop.'

'This was truly the first time I had ever really looked at *Lusitania* so thoroughly,' Moeller says. 'I've spent so much time over the years just focusing on *Titanic* and her sisters' design, that when looking into *Lusitania*, it really made me appreciate just how distinctive and unique this ship really was! From the themes in her plasterwork, the incredible amount of detail in the smallest carvings, and even down to her machinery, all was worn proudly on her sleeves.'

'It was always very important to me to get anything I worked on as close to the original as possible. Even with limited information we have on some areas, with the help from our incredibly generous historians, I always felt I was guided in the right direction even in the unknowns of this incredible ship. It has been challenging, but so incredibly rewarding, to have seen the final product; to see the work everyone had done with great pride come together was such an exciting sight! I'm extremely proud of what our team was able to accomplish and I'm forever grateful to have been a part of it!'

In a way, this was also a family project. While Tom and Emma Lynskey's son was en route, mother-to-be Emma delved into the tricky task of recreating many of the fabric textures and patterns from period photographs, as well as bringing some *of Lusitania*'s artwork to life. She says: 'While I thoroughly enjoyed recreating the various textiles and paintings aboard the ship, I found the stained-glass skylights to be the most exciting (as well as entertaining) work. It was endlessly fun to bring to life the charming and sometimes bizarre scenes originally contrived by the cherub-obsessed minds of the Edwardians. One of my favourite pieces is the skylight in the Smoking Room. Because the scenes in this skylight were barely visible in any known photographic documentation, I was able to include some of my own ideas (in keeping with the cherubic theme that pervaded the decor, of course) and depict three scenes from a favourite Victorian children's novel of mine. I'm very curious to find out whether anyone who plays the experience will recognise what story they are from!'

There was stained glass, frosted privacy glass pane patterns, furniture, carpeting and wood panelling – each piece of which was, in turn, handed off to lead animator and modeller Levi Rourke for him to 'plug them in' to the model. The most laborious, by far, were the oil paintings in First Class that needed to be painted in the real world, on wood, and scanned into the digital environment. A close second were the nearly two-dozen skylights that first needed to be deciphered from grainy, poorly framed photographs before their digital recreation.

A piano technician when not writing and researching ships, Layton even advised reconstruction of the rather unusual 6ft 2in Broadwood grand piano from *Lusitania*'s First Class Lounge and Music Room. Meanwhile, his wife Tessa joined the project by helping transcribe and record period sheet music about *Lusitania* that would feature in the museum experience. 'It was amazing to finally hear this music after we spent all the time transcribing and recording it.' Tessa says. 'My favourite piece was without a doubt "*Lusitania* Waltzes", by Edwin Goffe, written in 1910. It feels like there's many songs inside the one, with different feelings in each; it's really dynamic, and almost sounds like Beethoven.'

Meanwhile, Rourke had begun to rebuild the ship. 'It was a monumental task, no one had ever done anything to this scale really before. It included combing over the entire ship, and part of that meant rebuilding the hull texture and remapping it to have it fully complete. It took three weeks of hard work but resulted in a stunning *Lusitania*.' In a departure from previous projects that they had worked on, the exterior hull was modelled as one unit with the interior of the model, so that the ship works together as a single piece rather than being two separate models, an interior one and an exterior one.

When all was said and done, *Lusitania* had well and truly come back to life for the first time since 1915. Using the model, the team was not only able to release the virtual museum experience *Lusitania: The Greyhound's Wake*, featuring a real-time animation of the sinking; they were able to add their digital work into this permanent analogue format to preserve the memory of the great ship, and those who sailed on her, for future generations of enthusiasts.

When *Lusitania* was being designed, Cunard had given careful consideration to the idea of firing the ship's boilers with oil fuel, rather than coal. Naval architect Sir William White later said that Cunard 'recognised that by the use of oil there would be a considerable gain in weight and space required for fuel, large economies of labour in the stokehold and the numbers in its staff, easier regulation of steam, and more uniform production.'[9] It would also be much easier to refuel the ship while she was in port, as moving coal aboard a liner was a notoriously grimy, labour-intensive, loud and messy task. Despite these obvious advantages, Cunard could not obtain satisfactory arrangements for the regular supply of oil fuel at an acceptable price; coal was still the standard of the time, and would remain such until after the First World War, so it was decided to use coal. Converting *Lusitania* and *Mauretania* to oil later on was still possible, and this was eventually performed on *Mauretania*.

Once *Lusitania*'s powerplant and machinery had been installed, her superstructure could be built, and the interior spaces could also be partitioned off and turned into finished cabins and public rooms. Her final floor plan would primarily be laid out with yellow pine wooden walls dividing the cavernous, unfinished decks. Once this framing had been finished, each subdivision would slowly be transformed into a completed reality.

Lusitania's passenger accommodations were unquestionably the most spacious and beautiful of their time. Although she was being funded on a Government subsidy, Cunard understood the need to create stunning, extremely comfortable spaces for their passengers. In January 1905, Scottish architect James Miller had written to Cunard to ask that he be considered to design the interior of one or both of the new liners. Cunard liked his previous work and his terms, and in late August advised him that he had the position for *Lusitania*.

James Miller was the son of George Miller, a farmer in the parish of Auchtergaven, Scotland. The Miller family moved to Cairnie, Forteviot, shortly after James' birth.[10] Rising from humble origins, he was educated at Perth Academy; in time he became an associate of the Royal Scottish Academy, and a Fellow of the Royal Institute of British Architects. Some of his other accomplishments included work on the Canadian Pavilion for the Glasgow International Exhibition of 1901, an extension to the Glasgow Central Station Hotel

9 Imperial tons were equal to 2,240lb, not modern tons of 2,000lb each. Sir William Arrol and Co., Ltd. is well known for building the 'Great Gantry' at the Harland & Wolff shipyards where *Olympic* and *Titanic* were later built.

10 *The New York Times*, 9 July 1908.

Workers leave *Lusitania* at the end of a long day's work. The liner's upper works are progressing well, and funnels and lifeboats are already in place. (Russ Willoughby Collection)

(completed in 1907, in collaboration with Donald Mathieson), the Turnberry Hotel in Ayrshire and the Prince of Wales Museum of Western India in Bombay (completed in 1915). By 1909, he had achieved the distinction of being named one of Glasgow's 'Who's Who' citizens. Today he is recognised as one of Scotland's greatest architects.[11]

Miller created a theme that was very light in colour. It made use of a lot of white, but never grew boring to the eye; the white was supported by gilt detailing, grey- and rose-coloured accents, and combined with the use of natural French-polished wood of the finest quality. Overall, the decor was based on Louis XVI and Georgian themes.

11 In *Engineering*, 19 July 1907, *Lusitania*'s architect was referred to as 'James Millar, R.S.A., F.R.I.B.A.' Some modern *Lusitania* books also spell his name Millar. At least one volume spells his name as Millar *and* Miller in the same paragraph, and other books spell his name Miller.

Meanwhile, a number of reference works on architecture, including numerous references to Scottish and Glasgow architecture, spell his name Miller. Included among this list is a more modern reference work, *C.R. Mackintosh: The Poetics of Workmanship (Essay in Art and Culture)*, by David Brett (Harvard University Press, 1992). In an article on his work for the Glasgow Exhibition, *The New York Times* of 12 May 1901 spelled his name Miller, referring to him as 'one of the foremost Scotch architects'. The architectural firm that he later ran with his son, George, was also spelled Miller & Son.

It seems that all of the confusion over the spelling of his name comes from an original spelling mistake in *Engineering*'s special number on the liner, and the spelling mistake has been oft repeated in *Lusitania* books ever since.

James Miller, a prominent Scottish architect, created a scene of interior decor for *Lusitania* that was considered 'the highest conception of artistic furnishing and internal decoration'. (By Courtesy of the Mitchell Library, Glasgow City Council)

Regarding *Lusitania*'s interior decor, *Engineering* magazine noted: 'The express Cunard liner *Lusitania* ... is not only a great step forward from the mechanical engineering standpoint, but marks the highest conception of artistic furnishing and internal decoration.'[12] *Lusitania*'s decor was, indeed, extremely successful without overwhelming passengers with bombast, a trap into which some German liners of the preceding decade had fallen.

Within the shipyard, as the set time for the maiden voyage began to near, things would have grown both enormously complex and frenzied. To add to the stress, regular inspections by Cunard representatives could, in a moment, halt all work on a certain area of the ship if it was not being done properly. There were walls to panel, carvings to make and mount, fittings and furnishings of all sorts to be carried aboard, wooden floors to be laid, India rubber tiles to affix, carpeting to unroll and lay carefully, elevator cages to install, thousands of light bulbs to mount. Each cabin needed coat hooks, a folding rack net for each berth, a brass door step, a white metal handhold over every upper berth, a brass mortice lock on each cabin door and much more.

The extensive plumbing system had to be installed, tested thoroughly and be absolutely leak-free. The ventilating systems needed to be installed, connected, tried and tweaked. Storm rails needed to be fitted in all the ship's miles of passages and companionways. The list of projects must have seemed endless.

Meanwhile, plans to accept *Lusitania* and *Mauretania* in Liverpool and New York were also proceeding, and these preparations were running up against the same deadline. Liverpool was both the registered home port and also the primary English terminus of *Lusitania* and *Mauretania*; this was a logical decision since it was also Cunard's base of operations. The Mersey River ran in a rather southerly direction inland from Liverpool Bay. On the east bank was Liverpool proper, and on the west were the towns of Wallasey and Birkenhead. Along the eastern side of the Mersey were a series of piers, including the Prince's Landing Stage, or Prince's Dock. This was where the big Liverpool-calling Atlantic liners tied up to discharge passengers and luggage and take on a fresh batch for the next outward-bound voyage.

The trouble was that the water alongside the Landing Stage was of an insufficient depth to accommodate the new Cunarders. The Mersey Dock Board managed to ensure that dredging operations were carried out prior to *Lusitania*'s arrival. This was no small task, encompassing over 20 square miles of riverbed and removing roughly 200,000 tons of material. Once finished, there was a depth of 36ft even in low water – sufficient to allow *Lusitania* and *Mauretania* to dock regardless of the tide. For times when the weather was too rough to tie up at the Landing Stage, or it was occupied by another liner, *Lusitania* could dock in the Sloyne – an out-of-the-main-traffic backwater.

Liverpool was not an ideal port for ships of *Lusitania*'s ilk, however. Far more often than could have been desired, poor weather did not allow the liner or her sister to tie up at the Landing Stage. Worse yet, the Liverpool Bar would not always give sufficient depth of water so the new liners could enter the Mersey and gain access to their home port; if they arrived when the tide was too low, they would have to wait until enough of the tide returned to allow them to cross.

12 *The New York Times*, 23 November 1907.

Two views from nearly identical angles showing work progressing. In the left photo, the lifeboats are swung out to allow maximum space for workers; men and materials are all around the Sun and Boat Decks. In the right photo, the ship is virtually complete. (Russ Willoughby Collection, left; Authors' Collection, right)

On top of this, the Sloyne had a strong current on the ebb tide of 4–5 knots. This would have an extraordinary effect on the enormous hulls of the new sister speedsters, forcing Cunard to order two new buoys, called North Cunard and South Cunard, for use in the Sloyne. The buoys were constructed by Messrs John Bellamy, Ltd, and supplied by Pintsch's Patent Lighting Company. When built, they were the largest buoys ever made, each with a 16ft diameter, weighing 16¾ tons, and attached to 16 fathoms of 4½in stud-link cable chain. The new ships could tie up at these buoys and re-coal there, even offloading passengers and cargo by ferry if necessary. Even so, there were occasions when the great *Lusitania* or her sister were swept along in the current despite having both main bow anchors in the riverbed; on such occasions, the liner would have to start her engines, raise anchors and head back to her original location before dropping anchor again. Coaling at the buoy was also dangerous in poor weather. Due to these disadvantages, White Star moved their primary English terminus to Southampton in the late spring of 1907. Cunard stuck with Liverpool until after the First World War, but eventually followed suit. Thus, during the entirety of her career, *Lusitania* ran out of Liverpool and returned there at the conclusion of each round trip.

On the other side of the Atlantic, in New York, similar preparations were necessary, but not all of them had been in reaction to *Lusitania*'s forthcoming arrival. Between Sandy Hook – the official entrance point into New York Harbour – and the Cunard piers in Lower Westside Manhattan, a new channel was being dredged to admit large liners to the port. For some time the older Main Ship–Bayside–Gedney Channel had been the primary access to New York Harbour. However, by the early 1880s, this passage was no longer sufficient for the ever-expanding size of seagoing vessels. New York City resident John Wolfe Ambrose led the way in trying to improve the port's facilities; eventually it became an eighteen-year project that was not completely finished by the time he died. Existing channels were also deepened and improved. One of Ambrose's visions, however, was the creation of a first-rate deep-draught entrance to the harbour from Sandy Hook, the official end-point for west-bound passages. The going was slow, but by May 1899 – the same month that saw John Ambrose pass away – the US Government gave a contract to Andrew Onderdonk. In the 1901–02 session of Congress, a bill was passed naming the as-yet-uncompleted channel in honour of Ambrose. The newly named Ambrose Channel project was slated to cost over US$4 million and involved the removal of some 42 million cubic yards of muck from the bay. The project had initially progressed slower than expected.

The entire project was not officially completed until 17 April 1914, but Cunard hoped that it was finished enough so that *Lusitania* could use it on her maiden voyage in September 1907. On 2 August, the War Department told

Cunard's New York men – a group headed by Manager Vernon H. Brown – that *Lusitania* could use the Ambrose Channel on her way in, albeit at her own risk. It was expected that the new giant would draw 32ft of water on arrival; although other large liners had left New York drawing that much water via the older channel, Cunard still wanted to bring *Lusitania* in via the Ambrose Channel. For one thing, it was a much straighter cut, and for another, the new channel's current depth of 35–40ft would, it was thought, allow for a larger margin for error. However, the course was not yet buoyed, and no one was sure that the necessary markers would be in place by then.

Wanting to test the channel before *Lusitania* arrived, Cunard sent *Caronia* out via the new channel when she left on Tuesday, 27 August 1907. Permission was obtained the day before, and as this 'dry run' got under way, two leadsmen were kept busy taking soundings to warn of any hidden obstacles. The atmosphere was clearly tense for those working to take the ship out, but the passengers thought that it was exciting to be aboard for this historic voyage; they watched with a notable sense of superiority as the inbound *Kaiser Wilhelm II* wound her way about the twisting curves of the older channel.

Captain James B. Watt, *Lusitania*'s first Captain, arrived at John Brown during the spring of 1907 so that he could take time to familiarise himself with the ship before taking her out. (Authors' Collection)

In the end, the passage was successful, with Harbor Pilot Cramer claiming to have shaved a full half-hour off the trip to the open sea. However, there was still one question; *Caronia* was only drawing 30ft forward and 30ft 3in aft at the time of her departure ... would *Lusitania*'s draught be too great at the time of her arrival to safely use the unfinished Ambrose Channel? As *Caronia* steamed into the North Atlantic, and the dredging equipment resumed their task in her wake, the only certainty was that time would tell. Work on the channel would continue right up until the morning that *Lusitania* arrived. The final decision to take her in by that route was only made on Monday, 9 September, while she was in the middle of the Atlantic on her way to New York.

Even once *Lusitania* had entered the Upper Bay successfully and found her way towards the Lower West Side Manhattan piers that were used by Cunard, facilities were not ideal. Work on extending Pier 54, where she was to tie up, was freshly completed in time, yet it remained no more than a flat extension of land into the North River. There was no decent structure atop it to accommodate passengers as they arrived. Although Cunard put up a small temporary shed, there was no question that it simply would not hold all of the passengers once they had stepped ashore. Company officials could only hope that it didn't rain. The main pier structure would not be finished until very late in 1908, and in the meantime, *Lusitania* would stretch capacity to the limit.

Meanwhile, back on the Clyde, things were progressing rapidly. In the spring of 1907, a new figure was introduced to *Lusitania*. As spring turned to summer, he could be seen familiarising himself with what was, in reality, 'the first of a new class of giant passenger carriers'. This was Captain James B. Watt, *Lusitania*'s first Commander. Watt had just transferred from *Carmania*, a vessel that had entered service with Cunard in December 1905, just a year and a half previously. When *Carmania* had entered service, she was the largest ship in Cunard's fleet, and boasted a service speed of 18 knots. To Watt, as he became familiar with the new *Lusitania*, *Carmania* must have seemed a toy – she was only two thirds the size of, and was 7 knots slower than, the new vessel. The responsibility of commanding such a new and unique vessel must not have been lost on Watt, but the excitement must also have been palpable. On 31 August 1907 it was reported:

> It is appropriate that the man who will command [the *Lusitania*] is not only the commodore captain of the Cunard fleet, but we believe the commodore of the North Atlantic passenger trade … The responsibility of handling this great ship is of course an exceedingly grave one, inasmuch as her size and speed are beyond anything which a shipmaster has had the opportunity hitherto of becoming conversant with. Captain Watt, however, brings to bear upon his command a wealth of experience and skill that may be claimed to be unmatched. He is a type of commander which has made the British mercantile service honoured and admired throughout the world, and created almost sublime confidence amongst travellers.
>
> Like many other successful shipmasters sailing out of Liverpool, Captain Watt is a Scotchman. He is a native of Montrose, and comes of an old seafaring stock. His early days were spent in sail, and he had a thorough experience in windjammers on the North Atlantic trade. He joined the Cunard Line as a junior officer in 1873, and has passed through every grade and through most of the principal ships of the company, including the *Umbria*, *Etruria*, *Lucania*, and *Campania*.[13]

13 Period Cunard Line brochure.

The fitting out is progressing well: all the lifeboats are in place and funnels installed. Construction on the First Class Smoke Room and Verandah Café still lags behind, however. (J&C McCutcheon Collection)

The shipyard was a hive of activity, and the early September deadline for her maiden voyage loomed heavily as spring turned into summer. It was said:

> Four immense funnels, two tall masts, and a great dark hull alive inside and out with men is all that the public has yet been allowed to see of the great new Cunarder *Lusitania*, which, during June, was receiving the finishing touches in the tidal basin of Messrs. John Brown and Co.'s shipyard on the Clyde.
>
> By working day and night, and Sunday also, the builders of the newest, fastest and most luxurious ocean liner ever planned, hope to get the *Lusitania* ready to go down to the Tail of the Bank on June 27.
>
> Astonishing secrecy is being maintained in Glasgow and on the Clyde about the interior arrangements of this wonderful boat. The secrecy is ascribed in part to the fear lest rival companies should learn too much, and also to the fact that on the stocks near her lies the new turbine cruiser the *Inflexible*, to be launched on June 26, and now guarded day and night by armed men.[14]

More than simply finishing the ship in time, there was another sobering obstacle that her shipbuilders had to face

14 The passage, as well as the quotation in the preceding paragraph, was taken from the *Journal of Commerce* (apparently from Liverpool) of 31 August 1907. Our thanks to Watt's descendants Alec Watt and Michael and Elizabeth Gibson for their assistance and generosity in getting this right.

Smoke begins to drift from *Lusitania*'s forward funnels. She is preparing to venture from the nest for the first time. Not all the stokeholds have been fired, since the ship will be making the trip at a very cautious speed. (J&C McCutcheon Collection)

A rare view as *Lusitania* steams down the Clyde. Locals are clearly in evidence, crowding the pier on the right side of the photo, bidding Clydebank's greatest achievement a fond farewell. (Richard Smye Collection)

as well: getting the massive liner to the sea. It was said that the task would be tremendous, and that the dangers posed by the trip were 'a great weight of anxiety'.[15] The Clyde Trust and other river authorities had sent powerful dredging machines into the 14-mile passage for months, trying to ensure a minimum depth of 23ft at low water of ordinary spring tides. As the highest spring tide at Clydebank was never more than 11ft, the total depth of high water available under even the best of conditions would never have exceeded 34ft – very close to *Lusitania*'s ordinary draught. The date set for her to venture from her birthplace for the first time was Thursday, 27 June 1907, and the selection of that date was chosen to coincide with the highest spring tide, ensuring maximum depth of water under the vessel's keel. Since such favourable circumstances would not repeat for some time, it was crucial that *Lusitania* make it out of Clydebank on that day, or she would be stranded there for some time afterwards.

In order to minimise the ship's draught, only enough coal was put aboard to get her comfortably down to Gourock. When she departed, just before noon, she was drawing 29ft 9in aft and slightly less than that forward. Although it was several feet less than she would ordinarily draw, there were still concerns, particularly about dangers in the tight curves of the river. Fortunately conditions were favourable. The ship made the trek at a cautious speed of about 5 knots, with the assistance of a half-dozen tugs. Despite reportedly experiencing 'great difficulties' – to the point that some papers later reported incorrectly that she had actually gone aground – *Lusitania* made the trip unscathed. She also gave

15 *Star*, 20 August 1907, p.2.

Safe arrival: *Lusitania* has anchored at the Tail o' the Bank. Additional coal is already being loaded for her upcoming 'secret' builder's trials. (J&C McCutcheon Collection)

the first hint of her capabilities during the two-hour passage, for she 'answered her helm with more than usual promptness'.[16] With her first trip under her belt, she dropped anchor off Gourock, where she would spend the next few days. Although the official trials were not scheduled until the end of August, John Brown wanted to conduct a series of 'secret' trials to learn exactly what sort of ship they had built.

The first runs in open water, on Tuesday and Wednesday, 2–3 July, gave the first reason for optimism: she achieved a speed of 25 knots for a time. The ship's wake and the wave formation along her sides was also 'conspicuously small', indicating a well-formed hull, with 'remarkably low' resistance as she carved her way through the sea.[17] However, the tests were also a disappointment of sorts: significant vibration was manifest in the Second Class areas aft.

Returning to the Tail o' the Bank, John Brown carried out some stiffening of her structure to combat the vibration. When the ship once again took to open water at 5 p.m. on Monday, 15 July, she was run at 25 knots in the direction of Liverpool; the vibration astern showed a marked improvement, but not enough to prove satisfactory. John Brown reassured a nervous Cunard Line that they had more plans to combat vibration up their sleeves; among these would be the fitting of additional webs to the structure in the area of the forward propellers. Meanwhile, *Lusitania* crossed the Bar and entered Liverpool at 5 a.m. on the morning of Tuesday, 16 July. She anchored off the Landing Stage at Liverpool two hours later, entered the Huskisson deepwater dock later that day, and later proceeded into the Canada Dry Dock. While drydocked, her underwater hull was thoroughly cleaned and repainted, since it was 'heavily coated with the chemically saturated mud of the River Clyde' and other growths that had accumulated on it during the previous year. This would allow for an accurate and unimpeded set of results during her formal acceptance tests.[18] During this stay in drydock, her white-painted Shelter Deck hull strakes were also painted black straight out to the prow, ostensibly to make her appearance better match the appearance of other Cunard vessels.

On Monday, 22 July *Lusitania* was removed from drydock and returned to the Huskisson Dock. There, her bunkers were filled with coal – South Wales coal for the Nos. 1

16 *Ibid*.

17 *Engineering*, 12 July 1907. Papers that reported she had grounded included *Otago Witness*, 3 July 1907, p.69, among others.

18 *Engineering*, 12 July 1907.

Above: Lusitania's funnels and Boat Deck are picked out in soft light while she lies at the Tail o' the Bank. (HFX Studios)

Left: A splendid port-bow three-quarter view of *Lusitania* at the Tail o' the Bank. (J&C McCutcheon Collection)

This fine starboard view shows *Lusitania*'s appearance at the time of her builder's trials in July 1907: notice the white stripe at her prow that would soon be painted black. (HFX Studios)

2 July 1907: *Lusitania* is escorted by several tugs, moving towards the open sea for the first time. Smoke belches from her funnels as her stokers work up a strong head of steam for the first time. (Authors' Collection)

The Starting Platform in *Lusitania*'s Engine Room. It was here that her engineers awaited the command for 'Full Ahead' for the very first time. (Authors' Collection)

Based on a historic photograph likely taken in the evening of 15 July 1907, as *Lusitania* was steaming down to Liverpool for the first time, this artistic view shows *Lusitania* passing Ailsa Craig, a small promontory jutting from the Irish Sea just south of the Isle of Arran. (HFX Studios)

On the morning of 16 July 1907, *Lusitania* arrives in Liverpool for the first time. (Richard Smye Collection)

and 4 stokeholds and Yorkshire coal for the Nos. 2 and 3. Leaving Huskisson Dock with the morning tide, she departed Liverpool, passing the Rock Light at 12.24 p.m. bound for Scotland once again. She arrived there on Saturday, 27 July. Beginning at 4.40 a.m. that same day, the liner was run on the Skelmorlie measure mile seven times in each direction. Her draught was 32ft 9in, and she was displacing 37,080 tons. Her revolutions ran from 116.1 up to 194.3, and her speed ranged between 15.77 and 25.62 knots; at this latter speed, she was transmitting 76,000hp into the water. These results were satisfactory, indeed. She returned to the Tail o' the Bank that same Saturday afternoon and anchored.

Before the evening was out, *Lusitania* was to depart on a 'pleasure cruise' around Ireland. Aboard would be a specially selected group of about 200 VIP guests, 40 of whom were women. One of these was Mrs Ethel Tweedie, a Queens College graduate and authoress. She had a special interest in philanthropy and served on several committees of the International Council of Women; in five years, she would receive special recognition for rendering assistance to Sicilian earthquake victims. She later recalled of this trip:

> We left Euston at 10 a.m. [on Saturday] ... for the Clyde, in a saloon train with every comfort, for it was a 'Cunard special'. A tug quickly conveyed us on board from Gourock ... Representative people of all kinds were on board ... there were members of the Government, Naval Attachés from different countries,

On 17 July 1907, *Lusitania* was tied up at the Huskisson Dock. She would wait here until she could enter the nearby Canada Dry Dock. (Stuart Williamson Collection)

distinguished lawyers, admirals, engineers of eminence (naturally interested in the new boat), with a sprinkle of politicians, literary people, and leading lights of various kinds. Nothing could have been more jovial or interesting than the company.

What a monster that great four-funnelled vessel looked as we came alongside. She was far too tall to allow us to reach her top decks by a companion ladder, and consequently a door on the side of the vessel, on a level with our tug, admitted a ship's gangway, across which we merrily tripped.

Once inside, we were somewhere just above the water-line, and were promptly hurled by a lift to our own particular deck, and found our way to our own particular cabin as indicated on each passenger's ticket. Brass bedsteads and silk eiderdowns were also innovations in the shipping world.

Within a few minutes we heard the monster anchors weighed, and we were steaming as though it were an everyday occurrence, away down the Clyde in the evening light, past Arran and Kintyre, and out to sea. [Authors' note: According to one official report on

the ship's trials, the departure was at 6.55 p.m., and she passed Cloch Light at 9.10 p.m.] It was a glorious evening, the red sunset seemed to throw a halo of beauty over the first voyage of this great product of man's brain. It was as if a blessing smiled from that glorious sky ...

This trial trip was an emblem of construction and attainment of a great feat. An action of satisfaction and congratulations. All was joy ...

All Sunday we sped at twenty-two knots an hour, or three knots less than the colossal speed at which she was to cross the Atlantic. We saw the coast of Ireland clearly, although we were well out to sea ... By dinner-time on Sunday we were actually passing Queenstown, and by eleven the next morning we had arrived in the Mersey – that is to say, we accomplished some seven hundred and forty miles in thirty-six hours. The *Lusitania* could do a good deal more than that, as she afterwards proved ...

Somehow or other the *Lusitania* did not seem like a ship. Everything had been done to make her 'unshippy'.

Lusitania squeezing through the lock from the Huskisson Dock to the Canada Docks; the lock only gave 2½ft clearance for the ship's maximum breadth. Here the ship moves more or less north, parallel to the River Mersey; she would soon be turned to starboard to enter the Graving Dock. (J&C McCutcheon Collection)

Lusitania's sharp prow stands on the dock; she is clear of the water for the first time since her launch over a year before. Her lower hull gleams in fresh antifouling and a fresh coat of 'boot topping'. (J&C McCutcheon Collection)

The view from *Lusitania*'s forward funnel down over the Bridge and Forecastle while the ship was in the Canada Dock. (HFX Studios)

This view looking along *Lusitania*'s port side in the drydock shows the shoring timbers holding her in place. Her hull and funnels gleam in the sunshine. (HFX Studios)

Lusitania stands proudly above the dock, gleaming in all her glorious newness. Workmen would paint her C Deck plates black before she left the dock. (HFX Studios)

Lusitania leaves the Canada Dry Dock on Monday, 22 July 1907. Her prow has been painted black; the after pair of cranes have been removed, and one pair of stiffening stanchions have been installed astern to combat vibration. After this, she coaled at the Huskisson Dock prior to departing for Gourock. (Authors' Collection)

Right: This ultra-rare photo, taken during the Irish cruise, shows a group of women lounging on the Promenade Deck. From left to right: Lady Pirrie, wife of Harland & Wolff shipyard's Chairman, Lord Pirrie; Mrs Ethel Tweedie; Lady Inchcape; Lady Aberconway. (Authors' Collection)

Below: Mrs Tweedie's formal invitation for the Irish cruise. Once she had made a formal reply, the companies sent her a special 'Card of Admission' which would allow her to actually board the vessel. (Authors' Collection)

16.

NOT TRANSFERABLE

The Chairman & Directors of John Brown & Company Limited
and
The Chairman & Directors of The Cunard Steam Ship Company Limited
request the honour of the company of

Mrs Alec Tweedie

at the Trial Cruise (round Ireland to Liverpool) of the Quadruple Screw Turbine Steamer "Lusitania," on Saturday 27th inst. The Steamer will leave the Tail of the Bank at (about) 7.30 p.m., on arrival at Gourock of 6.30 p.m. train from Central Station, Glasgow, which connects with 10 a.m. train from London, (Euston).

Clydebank,
July, 1907.

Vessel will probably reach Liverpool early on Monday.

Card of Admission will be sent on receipt of acceptance.

An early reply is requested.
Addressed to Clydebank.

Cunard Daily Bulletin.

"Lusitania."

Vol 1., No. 1. Her Trial Cruise, July 28th, to 30th, 1907 Gourock to Liverpool
Distinguished Company aboard

Left Tail of the bank. 8-50 p.m., Saturday, Up to midnight, - - - - - 58 miles
Sunday - - - - - - - 488 miles
Monday - - - - - - - - 194 miles

Our First Impression.

Cunard enterprise keeps John Brown marching on. Hail, "Lusitania, the first new leviathan twin to wed the ocean: fleet, strong and luxurious, a challenge to the world's mercantile marine. May she have a long career, maintaining prestige and profit to the nation and her Company: (Editor).

A Really Modest M.P.

In reply to the constant enquiry
"Who is the Member for Sark?"

The Member for Sark
Still keeps in the dark,
No candle illumines his features;
But I have heard tell
Those who know him full well
Declare he's the best of God's creatures.
TOBY, M.P

THE KENNEL,
BARKS.
July 27th, 1907

Marconigram.

Capt. Dow, "Campania." 90 miles Westward of the Fastnet sent the following message to Capt. Watt, "Lusitania." "Wishing all on aboard a pleasant trip."

Loose-itania.

The directing genii: Wat-(&)-son.
Lucy Tania? A captivating ocean nymph
Kaid Maclean wires "Am singing 'Bule Lusitania"

Aid to appetite? A blow on the Promenade Deck.
Superb inertia. The Lounges.
"Comfort and courtesy," by Allison
"A luxurious shave" by Gadd
Complimentary to the cuisine "I'm off to dinner," Lord Pirrie.
Hoblyn's choice: A full saloon
To get the bane off, consult Beynon

A Good Beginning.

Why is the "Lusitania's" first achievement one that may well make all our distinguished statesmen envious? Because she has got round Ireland in one day

Unpardonable Omissions.

A vote of censure was passed on the builders and the Cunard Company alike, for one flagrant omission from the ship. She has neither a grouse moor, nor a deer forest aboard.

Atlantic Greetings.

On Sunday morning the "Lusitania" was in communication with Anchor Line "Caledonia," outward bound to New York, 170 miles west off Malin Head. Captain Baxter sent his greetings and good wishes to all on board.

The "Lusitania's" first Bulletin—all joy and happiness.

Lusitania steams down past Cloch Light in the vicinity of Gourock, on her way to open water for the Irish cruise. (HFX Studios)

Left: The very first edition of the *Cunard Daily Bulletin* printed aboard *Lusitania* contains a wealth of information about the goings-on aboard during the trip. The guests were reported, jokingly of course, to have found an 'unpardonable admission' from the ship's facilities. (Authors' Collection)

WELCOME ABOARD *LUSITANIA*

FIRST CLASS

Lusitania had seven First Class public rooms, five on Boat Deck (A), one smaller one forward on Promenade Deck (B), and another spanning Shelter (C) and Upper (D) Decks. We will begin our tour forward on the Boat Deck.

Writing Room and Library: Located just forward of the Entrance, the Writing Room and Library was in the Brothers Adam style. Wood panelling here, as elsewhere throughout *Lusitania*, was in an ivory-white colour applied in five coats. Panels of grey silk brocade were set into the walls. The carpet was rose-coloured, while a glass dome above the fireplace admitted light from above. An enormous mahogany bookcase on the forward wall contained a lending library; the books were wrapped in a special Cunard-branded leather binding. A number of writing desks were located throughout, each topped with 'a finely-chased mercury gilt lamp'. (HFX Studios)

Entrance: The Entrance gave First Class passengers access to any deck in their domain via the shortest possible route. This cutaway view shows how the Entrance and stairs ran from the Boat Deck all the way to E Deck. (HFX Studios)

The Boat Deck level of the Entrance, showing the stairs and elaborate black wrought-iron and gilt casing for the two electric lifts. The floor was laid in black-and-white India rubber squares. A skylight gives light from above. Across from the lifts on this deck, settees in rose-coloured upholstery complemented occasional tables and wicker chairs by a fireplace. (HFX Studios)

This view, taken between the D and E Deck landings of the Grand Staircase, shows how the stairs wrapped around the central lift tracks on three sides. The rose-coloured carpet continued from each landing down the stairs and each half-landing. (HFX Studios)

Lounge and Music Room: The late Georgian Lounge and Music Room was located just astern of the Entrance, between the Nos. 3 and 4 funnels, beneath the long skylight casing. This cutaway gives some idea of its location and layout. (HFX Studios)

The Lounge, seen here in a unique angle looking aft, was topped by an enormous barrel-vaulted plaster-and-stained-glass skylight. The skylight was subdivided into twelve panels featuring cherubic scenes, each based on a different month of the year. The carpet was green, while the furniture came in yellows and light-cream floral patterns. A Broadwood grand piano, veneered to match the room, featured prominently in the ship's concerts on each crossing. (HFX Studios)

Looking from port to starboard across the forward half of the Lounge, this view shows the French-polished mahogany walls inlaid with specially chosen, finely selected woods. Two enormous, working, marble fireplaces stood fore and aft; the green marble veining has been faithfully reproduced from the originals. A beautiful enamel painting hung above each mantle, one depicting wind and another the sea. (HFX Studios)

Smoking Room: From the Lounge, two sets of doors led aft to the Smoking Room, which was divided into two sections, a smaller one forward and a larger one aft, with a two-way fireplace set into the wall between. This view, looking to port, shows the forward section. Cabinets here contained items passengers could purchase, such as cigars, matchboxes and playing cards. (Authors' Collection)

Rendered in eighteenth-century style, featuring Italian walnut panelling and Queen Anne-style furniture, Miller's original design for the Smoking Room was rejected by Cunard on grounds that it was not outstanding enough. The final design, seen here in a view looking forward and to starboard, was truly magnificent. A large, barrel-vaulted, stained-glass skylight illuminated the space. All of *Lusitania*'s skylights were backlit at night, giving a comforting illumination no matter the time of day. (HFX Studios)

Verandah Café: Aft of the Smoking Room stood the Verandah Café, seen here at sea and in use in late June 1909. This space had undergone multiple revisions from its inception, and even when *Lusitania* entered service proved something of a disappointment. The aft wall was open to the elements, although by this time it could be enclosed with wooden screens; worse yet, the decor was bland, with hardwood seats and benches that made it look like nothing more exciting than an English train station. (Library of Congress, Prints & Photographs Division, Authors' Collection)

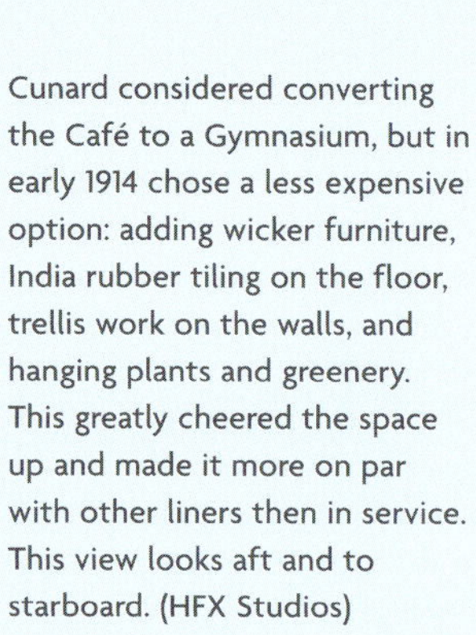

Cunard considered converting the Café to a Gymnasium, but in early 1914 chose a less expensive option: adding wicker furniture, India rubber tiling on the floor, trellis work on the walls, and hanging plants and greenery. This greatly cheered the space up and made it more on par with other liners then in service. This view looks aft and to starboard. (HFX Studios)

Observation Room: Forward on B Deck was the Observation Room, which overlooked the Forecastle through a quartet of portholes. Originally designed as an actual room, at the last minute Cunard decided to install two extra cabins, B1 and B2, reducing the size of the area and making it more of a glorified athwartship companionway than a room. (HFX Studios)

Dining Saloon: The focal point of all shipboard activity, the Louis XVI Dining Saloon – seen here in a stunning cutaway view – encompassed three decks. Seating for passengers was located on C and D Decks, while the elliptical dome extended up into B Deck. As originally built, the lower level could seat 321, and the upper level 143. (HFX Studios)

This view looks forward on the upper level of the Saloon. The dome featured four paintings in the François Boucher style, depicting the four seasons. The seats and Brussels carpet runners were in rose-coloured upholstery. Although seats were bolted to the deck, they could swivel for ease of entry and exit. An upright Broadwood piano was located on the upper level, and the band would play from this area each evening. (HFX Studios)

SECOND CLASS

Although Second Class public rooms were less spacious and ornate than those devoted to First Class, they were quite comfortable for the time. We will begin our tour on the Boat Deck.

It would be appropriate here to pause to consider a popularly held belief about *Lusitania* and compare it with facts. Although much has been made over the years of the use of plaster in the ship, one can only imagine the amount of maintenance that would have been required on large quantities of plaster installed on a vessel that was sent through the mountainous seas of the North Atlantic, where her hull would flex and twist under the stresses imposed on her 787ft length. In reality, most of the decoration in her First and Second Class public rooms was rendered in wooden panelling that was painted in an ivory white, to minimise the need for maintenance. Only a select number of items, such as the dome over the First Class Dining Saloon, were rendered in plaster.

Lounge: Located on the Boat Deck, the Lounge was panelled in French-polished mahogany, with a well in the middle surrounding a staircase that descended to the Second Class Entrance on the Promenade Deck. Above the well was a skylight. When vibration issues became apparent during the trials, John Brown was forced to insert enormous brackets and arches to stiffen the area; this view was taken before the arches were installed. (Authors' Collection)

Ladies' Drawing Room: Located forward of the Entrance on B Deck, the Ladies' Drawing Room had a dome that peaked up through the Boat Deck to admit light. Decor here was light, with a rose-coloured Brussels carpet and writing desks; chairs and banquettes were upholstered in finest horsehair. This photo shows the space populated with men, and may have been taken during the liner's early trials before entering service. Later, a large settee was placed in the centre of the room to help combat vibration. A cottage upright piano by Broadwood was located here. (Authors' Collection)

Smoking Room: Aft of the Entrance on the Promenade Deck was the Smoking Room, furnished rather like the Lounge above. The Smoking Room had a barrel-vaulted skylight that rose through the Boat Deck above to admit light in daytime. Although it featured fine mahogany panelling, it did not attempt to disguise hardwood deck planks on the floor. For seating, occasional tables and chairs throughout supplemented upholstered benches. (Authors' Collection)

Dining Saloon: The Dining Saloon was located on D Deck, and could accommodate 259 passengers at each sitting. Above the centre of the room was a circular well, lined with an elaborate white balustrade on C Deck; a Broadwood upright piano was located on the upper level, which could also be used as an overflow dining area. (Authors' Collection)

THIRD CLASS

Third class public rooms were far more Spartan than those of the other two classes. Yet they were comfortable, clean and of a high standard; while light on the excess decorations, they did offer prospective Third Class passengers a more-than-decent trip between continents.

Much thought was given to the lavatories and baths for Third Class passengers; with two baths for men and two for women installed. Nevertheless, these facilities were well equipped and offered a higher standard of cleanliness than most Third Class passengers were accustomed to in daily life.

Dining Saloon: The Dining Saloon was located forward on D (Upper) Deck; it could seat only 332 passengers, a fraction of Third Class's 1,186-person capacity; meals were thus served in two sittings, and the Smoking and General Rooms above on C Deck could be repurposed as overflow auxiliary saloons if necessary. An upright piano was fitted in the Saloon; tabletops were of unvarnished canary pine. Revolving chairs were each 15in in diameter, set in rows with 22in between the centres of each seat. (Authors' Collection)

Smoking Room: Forward on the port side of C Deck was the Smoking Room; it could seat 107, and had hardwood chairs located around tables, with slatted benches along the outboard bulkhead. This view looks forward from the adjoining Pantry. (HFX Studios)

General Room: On the starboard side of C Deck, opposite the Smoking Room, was the General Room; slightly smaller than its counterpart, it could seat seventy-six, and was furnished nearly identically. This view looks aft from the forward end of the room towards the Pantry. (HFX Studios)

Promenade: On either side of C Deck, just astern the Smoking and General Rooms, was a semi-enclosed Promenade, featuring more slatted benches along the outer hull beneath the portholes. A Bostwick gate at the aft end separated it from the open-air, First Class Promenade just astern on either side. (HFX Studios)

CABINS AND STATEROOMS

When *Lusitania* entered service, her passenger accommodations were set up in this way:

First Class: capacity 552 passengers accommodated in 260 rooms: (72 three-berth, 150 two-berth and 36 one-berth); all located more or less amidships on A, B, D and E Decks.

Second Class: capacity 460 passengers accommodated in 145 rooms: (85 four-berth and 60 two-berth); all located astern on C, D and E Decks.

Third Class: capacity 1,186 passengers, accommodated in 302 rooms (4 eight-berth, 21 six-berth, 227 four-berth and 40 two-berth); all located forward on D, E and F Decks.

More First Class cabins were soon added on E Deck, forward of the Grand Entrance, since *Lusitania*'s First Class accommodations proved even more popular than had been expected and were often overbooked. The alterations necessitated the removal of the old Third Class cabins and their replacement with new First Class ones, as well as the relocation of a watertight bulkhead. Originally located at Frame 167 outboard, jogging forward to Frame 169 inboard, the new bulkhead installed in its place was located further forward at Frame 181.

The most remarkable First Class passenger cabins were on B Deck. Two Regal Suites, the premiere accommodation on the *Atlantic* in 1907, were fitted just forward of the First Class Promenades on either side. The port Regal Suite, cabins B48, B50, B52 and B54, comprised a dining room, drawing room, two bedrooms, and a private bath and WC. The starboard suite was a similar, though reversed, arrangement, with the cabins being numbered B47, B49, B51 and B53. If let as a complete suite, each could cost from $1,250 (US) out of season, to $1,750 in the intermediate season, all the way up to $2,250 in full season – prices that seem especially steep coming as quoted in 1912.[19] The minimum First Class fare, by comparison, was a mere $115 at the time of the maiden voyage. So expensive were these Regal Suites that only the very wealthy were able to book them, and frequently they went as individual cabins instead of as full suites.

First Class passengers such as Austin Partner were impressed with their accommodation. Sailing out of Liverpool the day after Christmas in 1908, he described his cabin to his wife. 'My quarters on the ship are really most superb. I have a large room with dressing table, chest, writing table, wardrobe, sofa and washbasin and a nice brass bedstead ... it is really most sumptuous and quite the best I have ever seen. I like this ship much better than the *Mauretania*. Everything seems much brighter and more cheerful.' Less than four years later, Partner was lost in the *Titanic* disaster.[20]

The *en suite* rooms were a slightly less expensive option. Each of these suites consisted of a bedroom and sitting room, with a private bath and WC. In 1907, *Lusitania* sported six of these suites (B65 and 67; B68 and 70; B75 and 77; B76 and 78; B85 and 87 and B86 and 88); as the ship's career continued, the original six were given the designation 'Parlour Suite', and four new suites, given the original *en suite* designation, were installed further aft (B89 and 91; B90 and 92; B105 and 109 and B104 and 110). Each Parlour Suite went for $700 out of season and up to $1,500 in season; to add a second person to each suite was an additional $100 to the base price.[21]

Opposite, top left: A port-side, First Class, *en suite* stateroom. The door leads into the second stateroom of the suite forward. (Authors' Collection)

Opposite, top right: Although basic compared to the luxe of First Class, this Third Class two-berth cabin is clean and comfortable, with access to a washstand and comfortable bedding. (Authors' Collection)

Opposite, bottom: This view shows the E Deck Entrance, looking forward and to port. The port-side gangway doors, typically concealed behind sliding pocket doors that matched the panelling here, are visible. On the right is the doorway leading forward to the additional E Deck First Class cabins added after the ship entered service. In 1907, there was no door here, and the rooms forward of the bulkhead belonged to Third Class. (HFX Studios)

19 A full description of these public rooms was made in the 19 July 1907 issue of *Engineering* magazine.

20 Private letter by passenger Austin Partner to his wife, Mike Poirier collection.

21 These prices come from a Cunard Line brochure of the era. The seasons were described as follows. Eastbound: Out of Season: Mid-August to end-March; Intermediate Season: April and early August; Full Season: May, June and July. Westbound: Out of Season: November through March; Intermediate Season: April through July; Full Season: August through October.

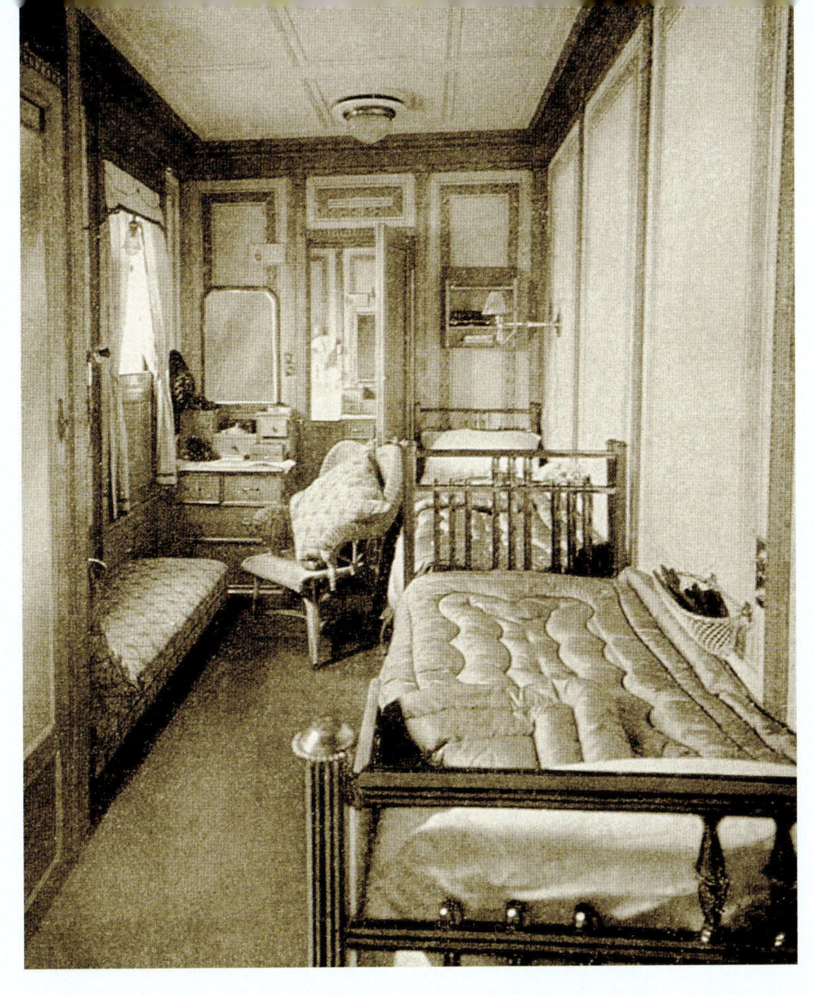

After detailing her reminiscences of the ship's beautiful public rooms, the marvels of the new vessel's elevators and how greatly 'appreciated' they were, Mrs Tweedie recalled of her travelling and dining companions:

> We were the merriest party, and our little centre table was not the least merry. It was circled by Sir Charles (the builder) and Lady McLaren, The Chairman and Mrs. Watson, Mary Lady Inverclyde, who christened the boat, Mr. Gerald Balfour, Lord and Lady Pirrie, 'Toby, M.P.' and Mrs. Lucy, the Hon. Charles Parsons, of turbine fame, and myself.

Indeed, she recalled: 'Nothing more sparkling can be imagined than the first glorious spell of rejoicing in the early days of the *Lusitania*, with a few hundred folk on board, and the air full of hope and pleasure of achievement.'

One of her fellow passengers, who was described only as 'a prominent British shipbuilder', was equally enthusiastic. 'Truly a magnificent vessel, not only in the perfection of her internal arrangements, furnishing and decoration, but the completeness of her structural detail is marvellous. I know nothing to equal her.'[22] *Engineering* magazine simply concluded that the results of the tests on the shakedown cruise were 'thoroughly satisfactory'.[23]

With the completion of the two-day, 740-mile 'pleasure cruise', *Lusitania* found herself at the Mersey Bar off Liverpool at 9.37 a.m. on Monday, 29 July 1907. There she sat, pristine in her freshly painted black hull and gleaming white superstructure, with her four majestic 'Cunard red'-and-black funnels towering over the whole assemblage. The Cunard ferry *Skirmisher* brought more guests out to the ship, including technical staff from Cunard and representatives of the Admiralty, all of whom were no doubt just as awed by her presence as the last party had been. She seemed to be moving even when she was at rest, when her reflection sparkled in the undisturbed water sitting by her hull. Her appearance denoted that of a racer. She seemed to throb with life and energy.

Once the fresh guests had embarked, and Mrs Ethel Tweedie and her party had disembarked – all via the tender *Skirmisher* – *Lusitania* set off to prove herself in a more official capacity. She departed at 1.18 p.m. and steamed towards the Clyde at a speed of 15¾ knots to make consumption tests. The speed trials began at midnight that night, and ended at about 1 a.m. on Thursday, 1 August; at the beginning, the ship's average draught was 32ft 7in, and when she finished it was 30ft 8in. This trial consisted of four runs over a measured course between the Corsewall Light on the Wigtown Coast and the Longship Light at Land's End. The weather was 'favourable' with bright sunny days, and nights just as brilliant. On both nights, however, there were stiff north-west winds running between Force 6 and Force 8. Since the ship was northbound on both nights, the wind had the effect of retarding her speed somewhat. On the other hand, everyone aboard must have been relieved because the wind prevented fog from rolling in and seriously affecting the tests.

The results of the four speed trials, with an accumulated distance of about 1,200 miles, were astounding. Southbound on the first run, she made an average speed of 26.4 knots, well above her design requirements. Northbound on the second, she averaged 24.3, a poorer showing but quite acceptable considering the interference from the north-west wind. The third run, with *Lusitania*'s prow pointed south again, produced a run only two minutes longer than the first run south, at an average of 26.3 knots. Coming back up again on the fourth, she made 24.6 knots, a third of a knot better than her first run north. The average speed for all four runs was thus 25.4 knots, almost a full knot better than the required average speed. To steam 1,200 miles, she had burned approximately 2,200 tons of coal. With all of these successes, John Brown must have breathed a tremendous sigh of relief.

However, *Lusitania* had more impressive figures to turn in before the trials ended. Once the ship returned to the Clyde on Thursday, 1 August, she made two additional runs between Corsewall Light and Chicken Rock at the southern tip of the Isle of Man. Over this 59-mile course, with a mean draught of 30ft 2in, displacing 33,770 tons, she averaged 26.7 knots southbound and 26.2 knots northbound, for an accumulated average of 26.45 knots, even better than her speed on the four long-distance runs. A half-dozen more runs followed between Holy Isle, on the east coast of Arran, and Ailsa Craig Light, which gave further proof of her high-speed capabilities and of her overall positive behaviour at sea. The weather on Friday, 2 August made it impossible to carry out the reversing and steering tests; instead, these would be carried out just prior to the maiden voyage.

22 *My Tablecloths: A Few Reminiscences*, Mrs Alec Tweedie, Hutchinson & Co, London, 1916, Chapter XXVI.
23 *Engineering*, 12/19 July 1907.

Under normal conditions, when the ship met the speed criteria called for in her builder's agreement, she would have been accepted by her new owners. However, there was still the proverbial fly in the ointment: vibration. This was not to be taken lightly; Second Class regions of the ship were still nearly uninhabitable. This proved unacceptable to Cunard, and the ship's formal approval was postponed. Further alterations were needed to reduce vibration. Naturally, the press and public alike were kept blissfully ignorant of this serious issue. Any rumours that did leak out were quickly minimised, particularly in maritime trade journals of the period like *The Shipbuilder* and *Engineering*.

With the clock ticking and the maiden voyage only a month away, there was no time to build and install new propellers, which were suspected of being at the heart of the problem. However, something needed to be done quickly to strengthen and reinforce the stern frame of the ship to reduce the effect of the vibration – otherwise the press reports surrounding the maiden voyage were sure to be marred, and the ship's reputation would almost certainly suffer damage right out of the gate.

Lusitania returned to the John Brown yards for some hasty, not to mention invasive, modifications. Numerous stanchions and arches were added to the Second Class interiors, bracing and stiffening the stern from within. Extra support columns concealed beneath Corinthian-style pillars were enough in some areas, like the Second Class Dining Saloon. Other areas needed more: the enormous new stiffening arches in the Second Class Lounge, for example, destroyed the original 'airy' feel of the space.

This extremely rare photo was likely taken shortly before *Lusitania* left the Clyde for Liverpool on 26 August. Although formal trials had been completed, Cunard would not formally accept her until 30 August. New stiffening arches are visible in Second Class areas astern. (Authors' Collection)

Meanwhile, financial problems were also creating tension. The original Government loan had totalled £2.6 million, to be divided evenly between *Lusitania* and *Mauretania*; funds were withdrawn monthly by Cunard and sent to each shipyard after they had submitted monthly accounts. In the end, construction costs on both ships ran significantly more than Cunard anticipated, exhausting the loan prematurely. The issue seems to have stemmed from the fact that these two ships were the most expensive ever built up to that time; there was no real precedent when the builders had submitted the tenders that won them the contracts. Additionally, the contracts had been set at a cost-plus, rather than a fixed, price, and had not included sums for the liners' lavish interior appointments. Cunard had also approved some extra, and costly, upgrades to their designs during construction.

It was a recipe for disaster, and if both ships met the contract specifications during their trials, Cunard had no legal leg to stand on in denying payment or rejecting the liners. Cunard, already in a bad financial position, was forced to raise capital on debentures to maintain their monthly payments. During 1908, they negotiated with John Brown for a rebate on *Lusitania*'s total price, which had skyrocketed to £1.65 million. John Brown granted a £30,000 reduction, meaning that Cunard paid £1,625,463 for her – some £325,463 over her half of the loan. However, as bad as things were with *Lusitania*, they paled in comparison to the cost overruns on *Mauretania*, and the hostile situation that developed between Cunard and Swan, Hunter on the matter.[24]

All of these financial problems were kept carefully out of the public eye. Meanwhile, once work on stiffening *Lusitania*'s hull had been completed, she left the Clyde bound for Liverpool on 26 August. Over the course of that day and the next, the opportunity was taken to carry out her stopping and steering tests.

With the ship's engines turning at 166 revolutions, pushing her forward at 22.8 knots, the officers on the Bridge rang down 'Full Speed Astern' on the Engine Room telegraphs; the engineers went to work quickly. The turbines were slowed and stopped in one minute and then reversed. In three minutes and fifty-five seconds the ship was brought to a complete stop, having moved forward about three quarters of a mile, or roughly six boat lengths.

The turning tests were also carried out at a speed just below 23 knots. First was a 'hard to port' order to the helm. The Quartermaster threw his weight into the wheel, and the tiller took some eighteen seconds to go over some 34½ degrees. The ship's stem shot to starboard, and she made a complete circle roughly 950 yards in diameter in five minutes and forty-eight seconds. When the tiller was put 'hard to starboard,' it took twenty seconds for it to travel 35½ degrees, again resting against its stops; the liner's prow shot to port this time, and she made a nearly identical circle in five minutes and fifty-three seconds. Astern half-circles were also made. On the next day, 27 August, the forward circles were made again, but this time at 180 revolutions per minute, roughly indicating 24 knots; she made the 'port' order in two minutes and forty-six seconds, and made the 'starboard' order in two minutes and thirty-five seconds. These results were hailed as a vindication of the decision, made during design, to cut away the 'deadwood' astern to aid her manoeuvrability.

During this two-day voyage, it was found that although the stiffening measures had helped the vibration issues, they had not entirely corrected them. Further corrective steps would have to be taken, but everything that could possibly have been done prior to the maiden voyage had been. Cunard formally accepted the ship, and she anchored in Liverpool's Sloyne; her official registration, numbered 124,082, was certified on 30 August 1907. With only a few days left before her maiden voyage, public excitement over the new British greyhound was at a feverish level. There was only one question on the minds of everyone: would she have what it took to reclaim British maritime prestige and maintain it?

Before proceeding with a discussion of the ship's maiden voyage and career, however, it would be appropriate at this point to discuss *Lusitania*'s technical features, to demonstrate just how unique she really was when her career began, and how she changed over the years.

24 For a complete breakdown on the issues surrounding *Mauretania*'s cost overruns and the resulting tension, please see *The Unseen Mauretania 1907: The Ship in Rare Illustrations* (2021, The History Press).

Opposite: A close view of *Lusitania*'s proud quartet of 'Cunard red'-and-black smokestacks. (HFX Studios)

3

LUSITANIA: A CLOSER LOOK

THE CUNARD COMPANY QUADRUPLE-SCREW ATLANTIC LINER

R.M.S. LUSITANIA

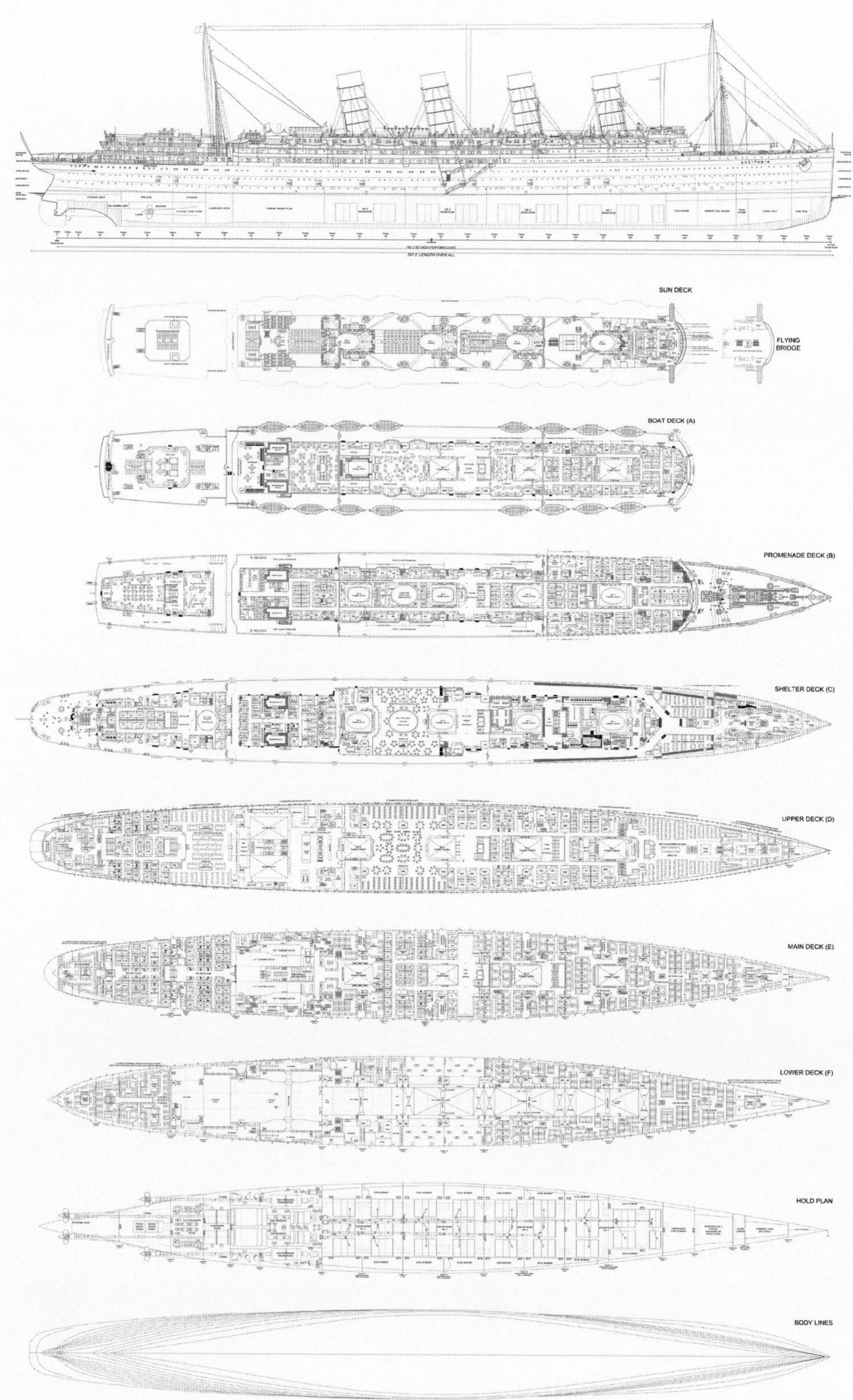

General arrangement plans of *Lusitania*, as she was originally built in 1907. (©2010/2023, J. Kent Layton)

SPECIFICATIONS

Specifications and Overview: Royal Mail Steamer *Lusitania*
Built by: Messrs John Brown & Co., Ltd
Laid Down: 17 August 1904
Launched: 7 June 1906
Service Career: 7 September 1907–7 May 1915

Overall length:	787.2ft	
Length between perpendiculars:	760ft 0in	
Width:	87ft 6in	
Draught:	33ft 6in	
Moulded depth:	60ft 4.5in	
Gross tonnage:	31,550.47 tons registered[1]	
Net tonnage:	8,514.59 tons registered [1907] 12,611 tons registered [1915][2]	
Displacement:	44,060 tons registered[3]	
Ditto/inch immersion:	112.4 tons[4]	
Powerplant:	Coal-burning Boilers (23 double-ended, 2 single-ended)	
Engines:	6 Marine Steam Turbines (4 ahead, 2 astern)	
Shaft horsepower:	68,000 nominal	
Screws:	4	
Number of decks:	10 (Sun, Boat [A], Promenade [B], Shelter [C], Upper [D], Main [E], Lower [F], Orlop, Lower Orlop,[5] Tank Top)	
Official number:	124,082	
John Brown & Co., Ltd hull number:	367	
Carrying capacity, maiden voyage:	Total carrying capacity, maiden voyage:	3,025
	First Class:	552 (260 rooms)
	Second Class:	460 (145 rooms)
	Third Class:	1,186 (302 rooms)
	Crew, designed:	827

1 This figure is not a determination of weight, per se, but rather a determination of enclosed space. 100 cubic feet = 1 ton. Gross tonnage is a measurement of the entire internal cubic capacity of a vessel.

2 Net tonnage is a measurement of the internal cubic capacity of a vessel remaining after the capacities of certain specified spaces (i.e., crew's quarters, machinery compartments and working spaces, etc.) have been deducted from the gross figure. The 1915 figure cited in testimony of Cunard's Assistant Superintending Engineer, Alexander Galbraith, Mersey Inquiry.

3 This figure was measured at a total equal to a quarter of the depth from the weather deck at the side (amidships) to the bottom of the keel. Displacement is the quantity of water displaced by a vessel when afloat. It exactly equals the weight of the vessel *and* whatever is on board at the time that the measurement is recorded. Displacement can be shown in cubic feet or tons: 1 cu ft of seawater weighs 64lb and 1 ton is equal to 35 cu ft of seawater.

4 It would take 112.4 tons of additional weight to lower *Lusitania* by 1in in draught. The approximate tons per inch of immersion (for salt water) is equal to the area of the waterplane in square feet divided by 420.

5 The Orlop and Lower Orlop Decks were only present at the extreme bow and stern sections of the ship; for the length of the Boiler, Engine and Auxiliary Machinery Rooms, the vertical space where they would have stood was left open to accommodate the ship's propulsive equipment.

MODEL TESTS AND CALCULATIONS: CREATING *LUSITANIA*'S BASIC STRUCTURE

Lusitania was 787ft in overall length, with a length between perpendiculars of 760ft; her extreme breadth was 87ft 6in, dimensions impressively greater than those of her predecessors. White Star's *Oceanic* (1899) was 685ft in length bp, while *Kaiser Wilhelm II* (1903) was 678ft in length bp. Because of *Lusitania*'s incredible dimensions, Cunard was unquestionably in uncharted waters.

The process of taking these initial figures and transforming them into *Lusitania*'s finished form was by no means arbitrary. Those involved in the task had to draw upon years of previous experience and analyse how different changes – even small ones – would alter the vessel's overall characteristics. Once the mathematical studies were complete and some of the best potential dimensions were arrived at, it was time to put them to the test through model studies.

John Brown, however, did not yet have a complete model-testing facility on their premises. Their model facility, then under construction, was unfinished; as there was no time to wait, they approached the Admiralty with the conundrum. Eventually they were able to begin model tests at Haslar in Hampshire, tests that *Mauretania*'s builders – Swan, Hunter & Wigham Richardson at Wallsend – also had access to and participated in.

Perhaps never before in the history of shipbuilding had such great pains been taken to analyse the best possible dimensions and form for a vessel. Different model shapes and sizes were tested to determine the best hull shape in matters such as efficiency in attaining the high speeds required, for stability, and for the design and positioning of the ship's propellers.

At the same time that the tests at Haslar were being carried out, John Brown had also managed to begin another set of model experiments; to accomplish this they had secured the cooperation of Messrs William Denny & Brothers, of Dumbarton. This firm, not far from John Brown, happened to have a finished model experiment tank. Later, after John Brown's own model experiment tank had been built, they ran all the tests again. These were done after most of the serious problems had been solved, but were able to provide further useful data, including the best possible directions to rotate the ship's propellers.

All the model experiments compiled a veritable mountain of data. One early proposal made by John Brown was for a vessel of 725ft bp by 80ft, with a displacement of 32,900 tons and a block coefficient of 631. After the model tests, this rough design proved woefully inadequate and was rejected. Tests representing vessels of 760ft bp and 80ft in beam and of the same length but 2ft narrower were also tested, but were also found wanting, particularly in stability. By increasing the models' represented beam to 88ft on the same length bp, satisfying stability was achieved. By next lengthening the ship's overall length to 787ft (in *Lusitania*'s case) or 790ft (in *Mauretania*'s), the necessary fineness of hull form could also be maintained. Extensive tests were also done to ascertain how much horsepower the different hull forms and sizes would need to drive them at the required 25 knots.

LUSITANIA AND *MAURETANIA*, COMPARATIVE LENGTH OVERALL

In the press of the period, and again in the years since her sinking, it has often been repeated that *Lusitania* had an overall length of 785ft.[6] At the other end of the spectrum, it was often stated, especially in period Cunard advertising that was repeated by some other publications, that *Lusitania* and *Mauretania* both had an overall length of 790ft. So what were the ships' true respective dimensions?

Mauretania's overall length was indeed 790ft, and a cursory glance at the length between perpendiculars for the two vessels, identical at 760ft, might initially lead one to believe that they also shared the same exact overall length.

Upon investigating the matter in a little more technical depth, however, *Lusitania*'s true finished measurements emerge:

From furthest point of ship aft to after perpendicular:	25ft
Length between perpendiculars:	760ft
From forward perpendicular to highest point of prow:	2.2ft

6 *Engineering* magazine, in its 1907 *Lusitania* special articles, referred to her as being 785ft long. *The Shipbuilder*, meanwhile, very specifically quoted her length at 790ft. Both journals referred to *Mauretania* as 790ft in length.

Lusitania's complete overall length was, then, 787.2ft, some 3ft shorter than *Mauretania*, even though their length between perpendiculars was identical. This measurement's accuracy is borne out by the fact that both ships were registered as being 762.2ft in length from the 'forepart of stem to sternpost head'. This measurement included the 760ft between perpendiculars as well as the 2.2ft of their prows, which extended forward from the waterline. *Mauretania*'s extra 3ft in length were entirely abaft the after perpendicular.

Interestingly, the 787ft dimension was by no means unknown at the time of *Lusitania*'s construction and entry into service. Two press reports stated:

> The *Mauretania* is a sister ship to the *Lusitania*, though somewhat longer than the latter vessel, being 790ft over all, as against the *Lusitania*'s 787ft, a fractional difference which, together with a trifling advantage in breadth, makes the *Mauretania* the largest vessel afloat.

This report was extremely detailed, even noting the slight 6in advantage in *Mauretania*'s width over that of her sister. A second report stated:

> The *Lusitania* is one of two ships built to the order of the Cunard Steamship Company to regain the blue ribband of the Atlantic ... Their length over all is 787ft, breadth 87ft 6in, depth 60ft 4½in, and draught 32ft 6in.[7]

Although the report was supposedly giving dimensions for both vessels, it is clear from the specific measurements of width and moulded depth cited that they were extrapolating from a set of *Lusitania*'s statistics and applying it to both vessels.[8] Another journal made a similar mistake at the time of *Mauretania*'s entry into service, taking *Lusitania*'s overall length and applying it to *Mauretania*, stating she was '787½ feet long'. Even though the figure was incorrect as applied to *Mauretania*, they clearly were not simply pulling '787 feet' out of the ether. Finally, The *Daily Mirror* reported:

> The *Mauretania* is the largest, fastest, and most comfortable passenger vessel afloat. She is about three feet longer than her sister-ship, the *Lusitania* ...[9]

When one compares the two ships' true dimensions, it is clear that *Mauretania* held all the titles of dimension and measurement between the two. She was 2.8ft longer, 6in wider, 387.22 tons greater in gross registered tonnage, and she also displaced 580 tons more than did *Lusitania*. However, as *Lusitania* was the first of the two to enter service, she alone held the title of 'the world's largest ship' until *Mauretania*'s maiden voyage.

STRUCTURAL MEMBERS, HULL STRENGTH, RIVETING

Lusitania really was in a whole new dimension of engineering territory. Her hull thus needed to be unusually strong in order to cope with the enormous stresses on the North Atlantic. Additionally, as a merchant vessel that also had to be strong enough for Government service in time of war as an armed auxiliary cruiser, *Lusitania*'s hull strength needed to be absolutely exceptional. This strength was achieved in several ways.

For starters, her frames were spaced closer together than typical merchant vessels of the time, some 32in apart amidships, narrowing to 25in aft and 26in forward. This required that the holes in the intermediate longitudinals, designed to lighten the keel's overall weight, were placed with their larger dimension running vertically instead of horizontally.

The cellular double bottom ran for almost the entire length of the ship, from the forward end of Cargo Hold No. 1 to the aft end of the Baggage Compartment behind the Auxiliary Machinery Room; its depth was 5ft beneath the Boiler Rooms and holds, and 6ft underneath the Engine

7 The draught measurement given in this instance is incorrect for both vessels; both were registered with a draught of 33ft 6in.

8 *Mauretania*, by comparison, was 88ft 0in in extreme breadth, and had a moulded depth of 60ft 3in and a 'depth to shelter deck' measurement of 60ft 6in. According to the registry details of both ships, *Lusitania*'s 'depth from top of beam amidships to top of keel' was 61.72ft, while the same measurement for *Mauretania* was 62.23ft. All of this is included to show that there were variations in their depth and width measurements, which demonstrates that the above citation was taking *Lusitania*'s dimensions and mistakenly applying them to both ships.

9 *Poverty Bay Herald*, 12 May 1906, p.6; 7 November 1907, p.5; *Otago Witness*, 20 November 1907, p.64; *Star*, 7 December 1907, p.4; *Wanganui Herald*, 31 July 1907, p.5; The *Daily Mirror*, 18 November 1907, p.3. As recently as 16 August 1993, the *Pittsburgh Post-Gazette*, p.6, referred to *Lusitania* as the '787-foot passenger liner'.

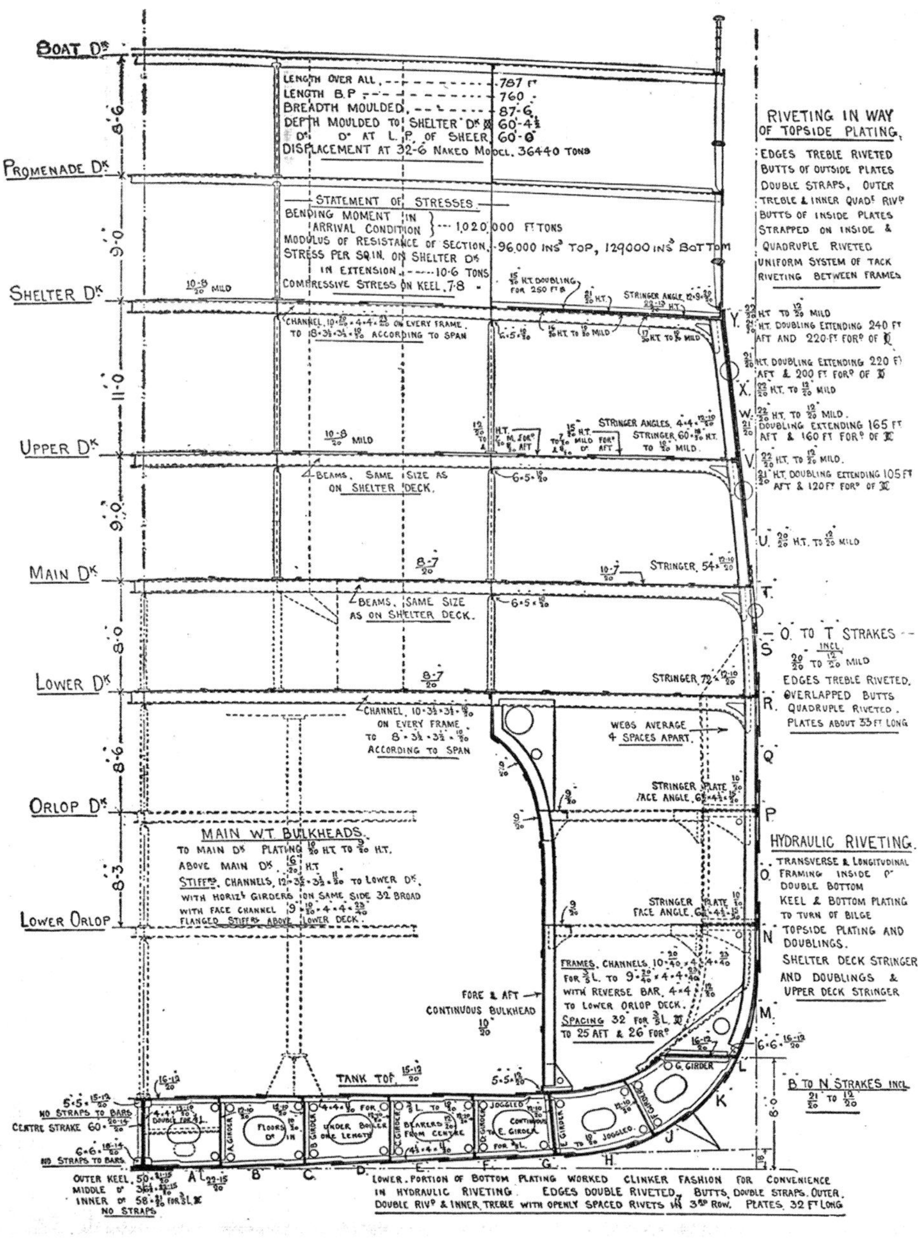

This midship section of *Lusitania* shows how her hull plating strakes were laid out, the dimensions of her structural members, details of riveting and many other fascinating details. (Bruce Beveridge Collection)

Room. It was subdivided by four longitudinal girders, two on either side of the keel, the second of which also capped off the top of the double bottom just past the turn of the bilge.

To aid in stability and comfort while at sea, a 36in-deep bilge keel was added to each side of the ship; it ran for just less than 230ft. These were roughly centred along the fore and aft length of the hull, flanking Boiler Rooms Nos. 2–4.

The plating that formed the Tank Top, otherwise called the 'inner bottom', was generally $\frac{1}{2}$in thick. The plates' thickness was slightly increased to $\frac{9}{10}$in underneath the supports that would bear the boilers' weight in due course. Underneath the seating for the turbines, additional stiffening girders were built in.

Stresses imposed on vessels like *Lusitania* as they moved through the sea were astronomical, especially in bad weather. Perhaps the worst-case scenario for *Lusitania* would be encountering a strong sea with wavelengths equal to her length while in 'arrival condition' with her coal bunkers depleted. Under this set of circumstances, not at all unheard of on the Atlantic, she would be the most prone to 'hogging', the term used to describe that moment when the ship's centre was supported on the crest of a wave and her ends were unsupported in the troughs. When *Lusitania* was 'hogging' in this scenario, the stresses on her uppermost structural deck would equal 10.6 tons per square inch (psi) amidships, while the compressive forces imposed on her keel would be 7.8 tons psi.

To aid in combating this stress, the idea of incorporating a higher grade of steel in the ship's hull plates was considered. Tests were carried out to determine whether high-tensile steel would actually be of use in strengthening the hull. The results showed that in general high-tensile steel would be 36 per cent better than normal mild steel, which was more commonly used. It was decided to employ this type of steel over a great portion of the upper hull. This allowed the builders to reduce the scantlings slightly to save top weight; however, they reduced them by only 10 per cent, leaving the hull of the ship considerably stronger than any previously seen in an ocean liner.

Typically, *Lusitania*'s hull plates were 32–3ft long. Amidships, the high-tensile steel plates were some $\frac{22}{20}$in thick. Towards the extreme bow and stern, where the frame spacing narrowed and the hull's form grew stronger, the plating was reduced to mild steel some $\frac{12}{20}$in thick. Along vast stretches of the hull's upper midships section – at D and E Decks – it was again strengthened through 'doubling', or the placing of one plate over another in high-stress areas of the hull. These high-tensile doubler plates were some 21⁄20in thick, making the total thickness of the shell plating in these areas 43⁄20in, an astoundingly strong design.

Holding the plates together were just over 1,000 tons of rivets. Great care was taken to ensure maximum strength at these connection points. The plans specified for *Lusitania*'s topside plating: 'Edges treble riveted butts of outside plates double straps. Outer treble & inner quadruple riveted butts of inside plates strapped on inside & quadruple riveted uniform system of tack riveting between frames.' From E Deck to approximately 10ft below the waterline, the edges were to be 'treble riveted, overlapped butts quadruple riveted'. For the plating of the lower hull, plates were 'worked clinker fashion for convenience in hydraulic riveting. Edges double riveted. Butts, double straps, outer double riveted & inner treble with openly spaced rivets in 3rd row.'

Interestingly, while all this attention was paid to the nature of riveting the plate seams, it was decided not to use high-tensile steel rivets. Common practice at the time was to use soft iron rivets in merchant vessels' hulls; the Royal Navy, on the other hand, used much stronger high-tensile steel rivets. It was thought that mild-steel rivets would suffice on *Lusitania*. Although exceeding industry standards, there was considerable controversy over the matter at the time. Yet the Admiralty did not press the matter, and the decision to go with mild-steel rivets stood.

Partly making up for this, most of the rivets were driven in hydraulically, after the rivet-holes were specially reamed to ensure proper seating and reduce stress points. Only where quarters were too cramped to allow the large, pincer-like hydraulic presses room to work were the rivets driven in the old-fashioned way.

SUPERSTRUCTURE AND EXPANSION JOINTS

The Shelter Deck was the topmost deck of the hull from the stern all the way forward to a point roughly 80ft forward of amidships, just behind the No. 2 funnel casing. From there forward to the tip of the prow, the hull was continued up to the Promenade (B) Deck. Everything above these points was only superstructure, which bore none of the ship's hull stresses. To save top weight, the superstructure was mainly built of lighter materials, aiding the ship's overall stability.

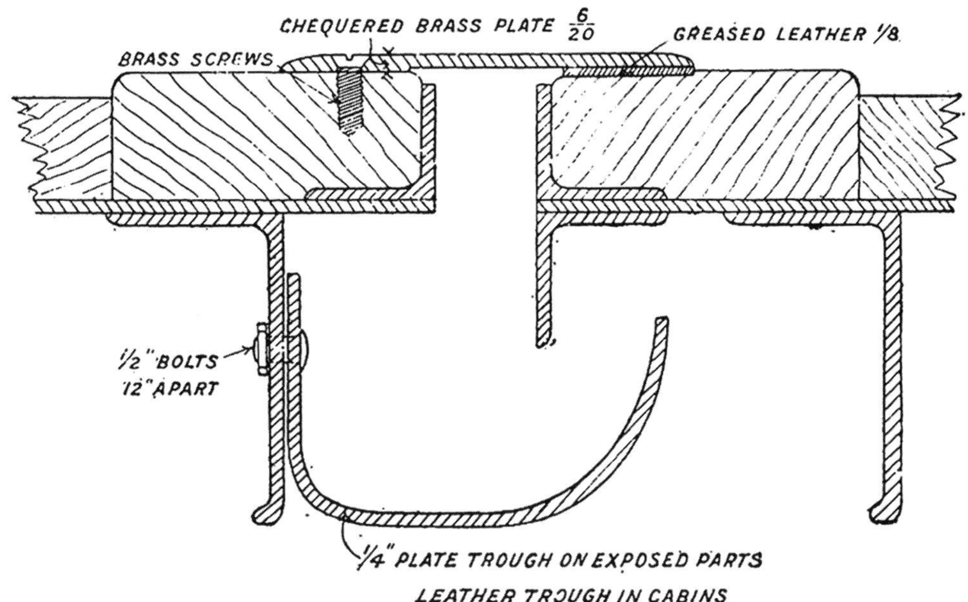

A plan showing how expansion joints were covered on the decks. On one side, a brass plate wide enough to cover the gap was screwed directly into the decking. On the other end, this plate was left mobile and unattached. To prevent noise as the plate moved, a small piece of greased leather was placed on the mobile side between the decking and the plate. Below the deck, a small plate trough was bolted into place on one end of the joint for exterior portions. Where the gap happened to cross through a passenger cabin, the plate trough was replaced with a leather trough. (Authors' Collection)

Interestingly, the original plan to have the Boat Deck (A) and Promenade Deck (B) overhang the rest of the hull by 20in on each side – an arrangement very similar to that finally adopted on *Mauretania* – was dropped on *Lusitania*. The Boat and Promenade Decks would be flush with the remainder of her superstructure. To prevent the superstructure from being torn apart by the flexing of the ship's hull as it worked its way through the sea, two expansion joints were introduced, one just forward of the No. 2 funnel casing, and one just astern of the No. 4 funnel casing.[10]

Astern of the primary superstructure was a secondary superstructure some 130ft long. This separate structure carried on the length of the Promenade and Boat Decks. Atop the Boat Deck was a smaller deckhouse with its own short segment of Sun Deck. This subsidiary superstructure housed Second Class cabins and public rooms. Connecting the two structures was a short bridge, which spanned the 5ft gap at the Boat Deck level. A locked gate and sign notified Second Class passengers that they were not allowed to cross that point into First Class portions of the ship.

10 *Mauretania* had the same two expansion joints cut into her superstructure. Interestingly, she also bore a third expansion joint, which was placed just forward of her No. 3 funnel. The designs and placement of expansion joints was far from an exact science at the time, and were introduced with varying degrees of success on any number of liners as they grew exponentially in size within just a few years.

AFT DOCKING BRIDGE

Atop the aft portion of the Second Class Boat Deck was the curved Aft Docking Bridge, the ends of which overhung the ship's sides by about 4½ft. From this vantage point, officers and crewmen could have a full view over the vessel's stern, sides and forward along the Boat Deck all the way to the forward Bridge Wing, a great advantage during docking manoeuvres. Underneath the centre of the Aft Docking Bridge was a structure that helped support the fixture's weight. Housed within was the Aft Wheelhouse, which housed a ship's wheel and third telemotor (two were fitted in the ship's Navigating Bridge, forward, and will be considered shortly); this allowed for a completely separate control over the rudder and hence the ship's course in the event that the primary Bridge equipment became unusable for some reason. Placing this backup inside the structure precluded the necessity for quartermasters to stand outside on the Aft Docking Bridge.

DECKS, REINFORCEMENTS, DECK PLANKING, CAMBER

Because the Shelter Deck was the top deck of the hull for the majority of its length, it needed to be particularly strong to protect against hogging stresses. It was also important that the hull's upper decks should be strong for another reason. In the event of war, if the Admiralty exercised the option of converting the liner into an armed auxiliary cruiser, she could be provided with twelve quick-firing 6in guns, four on the Forecastle (B) Deck, and the remaining eight on the Shelter (C) Deck.

To give some idea of the extent of this armament, a comparison to other First World War-era Royal Navy vessels puts things into perspective. HMS *Nottingham* (1914) was a light Town-Class cruiser of 5,440 tons. She sported only nine 6in guns – three *less* than *Lusitania*'s designed complement – with one 13lb anti-aircraft gun. The 605ft 9,800-ton cruiser HMS *Hawkins* (1919) had only nine 6in guns, plus four 4in anti-aircraft and four 3-pounder guns. This demonstrates the formidable level of armament planned for *Lusitania*, if she was ever converted into active military service.

Six-inch guns weighed from 5 to 7 tons each, adding up to 60–84 tons, depending on the exact specifications of the guns. All this weight had to be supported by the ship's structure. The decks also had to be capable of withstanding the guns' powerful recoil: each gun had a muzzle energy of over 5,000 foot-tons, and would be able to penetrate 4¾in armour at 5,000 yards' (2.8 miles) range, and 6in armour at 3,000 yards' (1.7 miles) range.

Specifics of the Shelter Deck's strengthening, where eight of the twelve guns would be installed, are telling. Amidships, stretching both fore and aft, the deck was not only plated with high-tensile steel, but was also doubled with a second layer of the material, ranging from 13/20in to 22/22in in critical areas. If installed, these guns would have been secured to the deck via circular mounts, or rings, which would allow them to rotate; such mounts were not originally built into the decks. At least some of these rings were installed aboard the ship in the spring of 1913; in the end, however, the guns themselves were never put aboard and *Lusitania* was not armed at any point in her career. To save top weight, the decks of the superstructure – Promenade (B) aft of the No. 2 funnel casing, Boat (A) and Sun – which bore none of the stress that the hull did, had far thinner plating.

The ship's outer decks were sheathed in yellow pine planking, with only the margin planks in teak. The exception to this was on the Forecastle, which was so exposed to the elements; here all the planking was teak. Many think that teak decking was the rule on Atlantic liners, but more often than not it was actually pine. Teak was far more expensive. All of *Lusitania*'s decking – teak or pine – was of the highest quality, free from sapwood, knots and other imperfections. Each strake of planking was 4in wide, and with a thickness of 2¾–3in. *Mauretania* and *Aquitania* also employed pine decking.[11]

Lusitania was designed as a 'wet' ship, a term that succinctly described her behaviour in rough weather. Her hull tapered to a fine point over a large portion of the ship's forward length. At the bow, as each deck climbed in succession out of the water, the widening 'flare' was quite fine by Atlantic liners' standards. This made the bow of the ship more or less slice through a large swell instead of attempting to climb over it in an exaggerated pitching motion. In rough weather, as she cut into rolling seas, it tended to send water straight up off the sides of the bow, thoroughly wetting the Forecastle. In particularly foul weather, the whole bow of the ship was often buried in solid water.

11 The ship's plans were quite clear on what kind of deck materials would be used in what areas. Many thanks to Bruce Beveridge and Scott Andrews for their assistance sorting out the pine vs teak deck planking standard on ships of the period.

Standard shipbuilding practice dictated building each deck with a camber, or upward crown, with their highest point amidships and the lowest points along the port and starboard edges. This allowed water that collected on the decks to wash out towards the edges of the ship, where it would drain away through scuppers. On a 'dry' ship, the camber, or crown, could be much less than on a 'wet' ship. The later White Star liners *Olympic* and *Titanic* were 'dry' ships, and sported a 3in camber for each of their decks. However, because large volumes of water would frequently wash over *Lusitania*'s open decks in dirty weather, her Boat, Promenade and Shelter decks were given an 18in camber, while a 6in camber was specified for all decks below that.

Along the centreline of the vessel, the height of each of *Lusitania*'s decks was as follows:

Promenade Deck:	8ft 6in
Shelter Deck:	9ft 0in
Upper Deck:	11ft 0in[12]
Main Deck:	9ft 0in
Lower Deck:	8ft 0in
Orlop Deck:	8ft 6in
Lower Orlop Deck:	8ft 3in

12 According to *Lusitania*'s original Builder's Specification Book, this deck was to be 10ft in height; by the time the ship had been constructed, however, this had been increased by 1ft.

The height of public rooms on the Boat Deck varied depending on their location, as the height of the deckhouse structure was elevated from a point just forward of the No. 2 funnel to the aft end of the First Class Smoking Room.

SMOKESTACKS, WHISTLES, SIRENS

Atop the Sun Deck stood *Lusitania*'s crowning, and most identifiable feature, a quartet of proud smokestacks, each endowed with a coating of traditional Cunard livery. These funnels would vent the exhaust gases from the boiler furnaces in the stokeholds. Each smokestack bore five segments. The lower four were painted in Cunard's unusual orange–red colour. Black bands separated each segment, and the topmost section was painted black. Each of the funnels was roughly 65ft tall;[13] being positioned atop the Sun Deck, they towered some 75ft above the Boat Deck. The funnels had a fore–aft diameter of 24ft, and were elliptical in shape, with their narrow dimension running port to starboard. Each structure was in fact a double funnel, with an inner casing separated from the outer

13 All four funnels measured approximately the same height from the Sun Deck to their peaks, a distance just shy of 65ft. Although the No. 1 funnel was mounted on a slightly lower portion of the Sun Deck, because of the ship's sheer line it was no shorter than the other funnels at the ship's designed load draught. As fuel and stores were consumed during a voyage, her draught would change, and if her bow rode higher than her stern, this would have tended to make the forward funnel the highest by a small margin.

This comparison shows how *Lusitania*'s whistles and sirens looked in their original configuration in 1908 (left) and in their final configuration in 1914 (right). (Library of Congress, Prints & Photographs Division, Authors' Collection)

casing by a minimum of 6in. The smokestacks were all raked aft at an angle of 10 degrees.

Ship funnels may appear to be rock-solid structures, but they are, in reality, made out of very thin steel and are quite delicate. Because of their relative structural weakness, each of *Lusitania*'s funnels was secured to the Sun and Boat Decks by a complex series of guy wires and chains for structural support.

The forward two funnels sported single-chime whistles in 1907, each one made of brass and standing 4ft high. The forward funnel's whistle was eventually replaced by a double-chime set-up to facilitate communication with tugboats, especially in New York Harbour, which was often quite noisy. Even this was found deficient, and was again upgraded to a triple-chime set that looked remarkably like the set-up sported by *Olympic* and *Titanic*. In addition, two sirens flanked the whistle(s) on the forward funnel. All of this equipment could be controlled from the Bridge by the captain and his officers.

Early models for the ship, which was then slated to be powered by reciprocating engines in a triple-screw configuration, showed *Lusitania* sporting only three funnels. However, when her engine configuration settled upon turbines and the final boiler arrangement was laid out, it was deemed necessary to give her four funnels. On *Lusitania*, each funnel was tied to its own Boiler Room, and all four were operating smokestacks. German greyhounds like *Deutschland* sported four funnels, so it sat well with Cunard management that *Lusitania*, *Mauretania*, and eventually their successor *Aquitania*, would all sport four, as well.

Interestingly, to all general appearances, *Lusitania*'s funnels were evenly spaced, a departure from the 'double-pair' design of previous German ships of the era. However, their spacing was not precise: the gap between the Nos. 3 and 4 funnels was over 2ft greater than the spacing between each funnel in the forward and aft pairs; the gap between the forward two funnels was also a few inches greater than the gap between the Nos. 3 and 4.

VENTILATORS

Around the base of the four funnels, all along the Sun Deck, were twenty-two hinge-topped ventilators, eleven on each side (two additional ventilators of this type sat astern atop the roof of the Second Class superstructure). These primarily provided ventilation for the lower compartments of the ship, particularly the extremely hot Boiler and Engine Rooms. They also served as an escape route to the Sun Deck for anyone trapped below in the machinery spaces during an emergency.

These ventilators were truly unique in the annals of maritime construction. They presented a streamlined appearance quite different from the standard cowl-type ventilators; their low profile also made the four funnels of the ship appear taller. They were also the most easily identifiable difference between *Lusitania* and *Mauretania*,

Two photos showing *Lusitania*'s original and later vent configurations, as seen in December 1912. The cowl vents had been installed since May, likely while the ship was laid up for turbine repairs just before this photo was taken. (Jim Kalafus Collection [upper], Library of Congress, Prints & Photographs Division, Authors' Collection [lower])

for the latter ship bore standard cowl ventilators. It is arguable that *Mauretania*'s appearance suffered somewhat from this arrangement, but the simple fact was that the standard cowl ventilators worked more efficiently. The canister-like lids of *Lusitania*'s ventilators were often nipped cleanly off their bases in bad weather; they were progressively replaced with the more standard cowl ventilators as the ship's career continued.

BRIDGE HOUSE, OFFICERS' QUARTERS, WHEELHOUSE AND INSTRUMENTS

Just forward of the No. 1 funnel, atop the Sun Deck, was the ship's Bridge House, situated 90ft above the keel and 58ft above the normal waterline. The rectangular structure sported a curved face towards the bow and was 50ft in length. In the stern section of this structure were quarters for the ship's senior officers and the Officers' Smoke Room, which was serviced by its own pantry. The Captain's Suite was housed directly underneath the Bridge, on the forward Boat Deck. It consisted of a bedroom, day cabin and private WC and bathroom. From there, the Captain could obtain quick access to the Bridge via an interior staircase leading up into the Bridge House. Forward of the Officers' Smoke Room and Officers' Quarters there was a corridor that ran across the Bridge House, with a door to the exterior on either side. Just in front of this corridor, on the port side, was the Chart Room; on the starboard side sat the officers' WC and bathroom. Lining the front of the Bridge structure, behind eleven rectangular windows overlooking the bow and Forecastle, was the main Bridge. This was the nerve centre of the ship.

Most of the Bridge had rubber matting covering the deck to prevent slippage. Just behind the forward windows, arranged in a neat line, stood all of the ship's main control

A cutaway showing *Lusitania*'s Bridge and Wheelhouse structure as they could never be seen in a physical photo. (HFX Studios)

Right: Lusitania's Bridge as it originally appeared in 1907. (Authors' Collection)

Below: A Duplex signal lamp indicator on *Mauretania*'s Bridge; *Lusitania*'s set-up was very similar. (Authors' Collection)

equipment. These instruments were built to a higher standard than was the norm in merchant vessels of the period. This added expense was justified by the Admiralty, not because of *Lusitania*'s unprecedented size, but instead because in time of war she might be taken over for active duty as a naval auxiliary.

There were four telegraphs to control the turbine engines. The outboard telegraphs controlled the low-pressure and astern turbines, while the inboard telegraphs controlled the high-pressure ahead turbines. There were three controls for the ship's whistles, two pillar-style, loud-speaking telephones (enabling the Officer of the Watch to speak to the Engine Room Starting Platform, Aft Docking Bridge, Crow's Nest and Forecastle Head), telltale indicators to show the operation of the turbine engines, a steering telegraph to show the position of the rudder and a control for the watertight doors. Amidships there was a binnacle housing a compass. In the port-aft quarter of the Bridge was a Pearson's Fire Indicator, which was connected to a system of automatic and 'break the glass and press'-type alarms. The Officer of the Watch could thus be immediately informed of a fire in any of the monitored areas of the ship. A similar indicator panel was also located in the Engine Room. In the starboard-aft quarter of the Bridge was the watertight door indicator, which kept the Officer of the Watch apprised of the position of each watertight door. The electrically powered indicating panel contained a faceplate with an engraved plan of the ship. Backlit ruby-coloured discs were inserted into the plate to represent each of the monitored watertight doors. When

Left: A close-up of some of *Lusitania*'s Bridge equipment, including (from left to right) the starboard high-pressure engine order telegraph, the Graham's Patent loud-speaking telephone, and the Stone-Lloyd watertight door controlling gear. (Syler Beaucage Collection)

Below: This advertisement for Brown's Patent Telemotors, as fitted aboard *Lusitania*, proudly proclaims they were also fitted aboard *Mauretania* and *Aquitania*. The company also supplied the telemotors for the *Olympic*-class liners from the White Star Line. (Jerry Davidson Collection)

the doors were opened, the discs were lit; when the doors were closed, the discs went dark.

Amidships there was a small enclosed Wheelhouse, with five forward-facing windows looking out into the Bridge. Within that Wheelhouse – accessed by a door to port and starboard, and with a third door aft leading into the Chart Room – stood the ship's wheel. There were two complete telemotors fitted in the Wheelhouse, which were operated from the central wheel. A clutch allowed each telemotor to be engaged separately; also located within the Wheelhouse was a locker for the ship's signal flags. A small patch of matting just behind the wheel added to the sheltered environment to make a relatively comfortable space for the duty Quartermaster to stand while he guided the ship on her course.

From the Bridge, sliding doors to port and to starboard led out to the Bridge Wings. Rubber matting lined the decking outside the Bridge House. Aft of the sliding doors were staircases on either side that gave access to the Bridge roof; there, two skylights allowed light and, when the weather was

Brown Bros. & Co.,
LTD.,
Rosebank Ironworks,
EDINBURGH.

Telegrams: "Hydraulic, Edinburgh." *Telephone* 16 Central.

ON ADMIRALTY LIST.

Manufacturers of Highest Class
Steam and Electric Hydraulic Steering Gears.

Steering Gears and Telemotors fitted on board the
Q.T.S.S. "AQUITANIA."
" "MAURETANIA,"
" "LUSITANIA,"
&c., &c.

Original Patentees of the
TELEMOTOR.

All the latest Battleships and Cruisers in the British Navy, as well as a large number for most of the Foreign Navies, have been fitted with Brown's Patent Telemotor.

HYDRAULIC CARGO INSTALLATIONS.

Right: The port low-pressure astern engine order telegraph stands in front of the arched bracket that controlled the ship's steam whistles. (Syler Beaucage Collection)

Far right: The starboard-side twin-docking telegraph, manufactured by A. Robinson & Co., standing just outside the starboard door of the Bridge. Inside, just visible beside the door into the Wheelhouse, is a tantalising glimpse of a frame displaying a large sheet of paper; what was printed on the paper cannot be discerned. (Syler Beaucage Collection)

good, fresh air into the Bridge House structure. There was a compass platform atop the Bridge, further removed from all metallic influence;[14] later in the ship's career, a second platform was installed a bit further aft, between the Nos. 1 and 2 funnels. Still later on, the compass structure atop the Bridge was raised to prevent magnetic interference. On the starboard side, just outside of the Bridge, was a docking telegraph, while a control for the ship's siren was mounted on the bulkhead of the Bridge, just forward of the door. On the opposite side there was a docking telegraph and an anchor and lookout telegraph. Another control for the ship's siren was located on the port bulkhead of the Bridge.

Stretching further outboard were the actual Bridge Wings, which extended over the Boat Deck below, and were supported by large columns at the outboard edge of that deck. Each wing overhung the side of the ship by about 4ft, giving excellent visibility fore, aft and along each side of the ship, as well as over the Sun and Boat Decks astern to keep tabs on important goings-on. Wooden gratings covered the decking of the actual Bridge Wings; they were divided into four segments along each wing. Just behind the wings, staircases led down, giving immediate access to the Boat Deck.

MARCONI SET AND MASTS

Astern of the Bridge, also atop the Sun Deck, was the ship's Marconi Shack. This closely resembled its name, being a basic rectangle in shape, and having all the appearance of an afterthought. It was situated between the Nos. 2 and 3 funnels directly abaft the skylight over the First Class entrance. Twin aerial wires ran down to it from the dual aerials suspended between the masts. Within this small cubicle was the all-important Marconi apparatus, which connected the liner with the outside world and with other vessels while

14 Cunard specified that all rails within a 10ft radius of any compass were to be of brass construction in order to prevent magnetic interference. All of the ship's rails along the Promenade, Shelter and Boat Decks were of four rails each, topped off with a teak rail. Each rod of these rails was ¾in in diameter. On the Forecastle, the rails were comprised of five rods each, with the top rod being 1½in in diameter, with the lower four being ⅞in in diameter. No teak top was fitted to the Forecastle rails. All the rails were galvanised.

The starboard side of the Bridge, with the Wheelhouse amidships. The door to the left of the Wheelhouse led astern to the Officers' Smoke Room and their quarters. The ship's watertight door indicator panel, visible at left, is replicated faithfully from an original photo. (HFX Studios)

she was at sea. The ship's daily newspaper, which would be published on board during each crossing, would keep passengers informed of important events going on in the world while they enjoyed their voyage.

There were actually two complete sets of wireless equipment, one for transmitting and the other for receiving. The receiving set was placed in a soundproof 'quiet' room in one corner of the Marconi Shack. The wireless set was powerful enough to maintain communication with land for the entire voyage, except for a stretch of about three hours in mid-ocean. There was also an emergency apparatus available so that during an emergency where the power was lost, wireless contact could still be maintained with nearby ships.

Two operators were employed by the Marconi Company to man the set, keeping a twenty-four-hour watch. This was the first time a round-the-clock watch had been set up on an ocean liner, for it would not become mandatory until after *Titanic*'s sinking in 1912. Typically, *Lusitania*'s operators would have two six-hour watches each day.

All Marconi-equipped ships or Marconi stations were given a unique three-letter call sign. Ships had call signs that began with the letter M, while fixed shore stations began with the letter G. Land's End station, for example, was 'GLD'. *Lusitania* was given the call sign 'MFA', while *Mauretania* was designated 'MGA'. Passengers wishing to send wireless messages could do so from the Enquiry Office. Charges varied greatly depending on the stations a message was transmitted through; all wireless messages had to be prepaid, and if the delivery failed, or needed to be re-routed through another station, the tolls were recomputed and collected from the sender. The Marconi Shack was connected with the rest of the ship via telephone on a twenty-three-extension system that did not require a switchboard exchange. Communication could thus be made between the operators and the Captain and officers, or with the First Class Bureau on B Deck, where passengers conducted necessary business with the ship's Purser and his staff and composed their wireless transmissions.

Supporting the twin aerials that served as the Marconi antennae were the Foremast and Mainmast. Each rose roughly 170ft above the waterline at their peak and, like the smokestacks, stood at an angle of roughly 10 degrees from

Lusitania's Marconi set, as it appeared in late 1907 or early 1908. From left to right can be seen the printer, the coherer, the telephone, the bell alarm, the electromagnetic receiver, the tuner, induction coil and key. (*Electrician and Mechanic*, February 1908; Authors' Collection)

the ship's vertical fore and aft. The Foremast rose from the Forecastle, just behind the crew stairwell that led down to the Shelter Deck. The mast itself was hollow, with a ladder inside, so that lookouts could climb to and from the Crow's Nest, the base of which was over 85ft above the waterline, while protected from the elements. Above the Crow's Nest was the ship's forward navigation lamp. Still higher, atop the very peak of the mast, there was an unusual weather vane so small that it is nearly invisible in most photos. Attached to the Foremast were the crane booms that could hoist cargo from dockside up over the open cargo hatches and down into the holds. About 500ft aft of the Foremast was the Mainmast. This feature rose from the Shelter Deck between the First and Second Class superstructures, and its base was sunk solidly down into the Main Deck for structural strength. There was a navigational light attached to this mast, as well.

The Purser's Enquiry Office on B Deck. Located within the First Class Entrance, this is where passengers would conduct shipboard business such as sending telegrams. (HFX Studios)

MANOEUVRABILITY AND STEERING

In order to help make *Lusitania* and *Mauretania* highly manoeuvrable, it was decided to remove the ships' keels for a distance of 60ft forward of the aft perpendicular, just forward of the rudder. This feature, referred to as 'removing the deadwood', created an arch in the hull that allowed *Lusitania* to make tighter manoeuvres at high speed than would otherwise have been achievable.

There was a concern that because of this feature, *Lusitania*'s stern might not be able to support itself during a drydocking procedure, as the after 84ft of the ship would be completely unsupported. Extensive calculations were performed to see how the stern would react under those circumstances, and how much pressure would be exerted upon the last keel block supporting the stern, the structural point that would act like a fulcrum for the stern's whole weight. Fortunately, it was discovered that her stern would be sufficiently strong to support itself under these scenarios.

The Admiralty specified that all of *Lusitania*'s rudder be placed below the waterline, to protect it from damage during enemy action. The rudder was of the balanced type, supported by a single gudgeon and with a removable pintle. It consisted of three separate castings connected by horizontal flanges. These flanges had grooves, or recesses, called rabbets cut from their edges to receive their opposite member; they were then heavily bolted. The rudder head was of forged steel; its total area was 420 sq. ft, and its weight was 126,336lb, or roughly 63 modern tons.[15] It could move through 35 degrees on either side of the ship's main longitudinal axis (70 degrees in total movement), and up to 37 degrees (74 degrees in total movement) before physically jamming against its stops. The combination of the 'deadwood' removal and the highly manoeuvrable rudder allowed *Lusitania* to perform impressive turning circles on her trials and she proved highly manoeuvrable during her career.

All machinery for the steering gear, which drove the rudder, was likewise placed below the waterline for protection against enemy shellfire. The steering engines were

15 As originally reported, the rudder weighed 56 tons 8 centreweight. Imperial tons were some 2,240lb, while a centreweight was 112lb.

mounted just forward of the tiller; there was a primary engine as well as an identical back-up. Each had the capacity to exert 1,240ft tons of pressure against the rudder-stock.

WATERTIGHT SUBDIVISION

Watertight subdivision of *Lusitania*'s hull was important during her peacetime career, but was of even more concern to the Admiralty if she ever saw active service with the Royal Navy.

Lusitania was subdivided into twelve primary compartments by eleven transverse (port to starboard running) bulkheads. Behind the Forepeak there were two cargo holds, a transverse coal bunker, four Boiler Rooms (numbered 1–4 from bow to stern), the Engine Room, the Condenser Room, the Auxiliary Machinery Room, and the Steering Gear Compartment. The main bulkheads were built entirely out of high-tensile steel. In their lowest portions, they were fully ½in thick, and then thinned to 9/20in up to the level of the Main Deck; above that level, they were 6/20in thick. They were strengthened on one side by vertical channel stiffeners to the height of the Lower Deck, and by flanged stiffeners above that. They also had horizontal strengthening girders on the same side as the vertical stiffeners.

However, the Admiralty wanted even further watertight subdivision in case she ever came under enemy fire as an armed auxiliary cruiser. Normal naval combat vessels were purpose-built for such action, and sported heavy steel armour belts to prevent their flanks from enemy shells. However, this sort of armour plating was heavy and slowed a ship's maximum speed. As a passenger liner, intended to be the world's fastest, *Lusitania* could not be burdened with such armour. Yet without it, one well-placed enemy shell could penetrate her vital Boiler and Engine Rooms and wreak havoc on her machinery. In the Boiler Rooms, this could mean that boilers or pressurised steam lines could explode; in the Engine Room, the turbines could easily be damaged beyond repair. In either case, it would be quite likely that her entire powerplant could go down. This would force a drydock overhaul that could last months, or even jeopardise the ship's overall safety.

Some sort of protection had to be afforded the essential machinery spaces, so the Admiralty insisted that all vital equipment be placed below the waterline. Furthermore, they mandated that the Boiler and Engine Rooms be lined with longitudinal (bow to stern running) bulkheads in addition to the transverse ones. These longitudinal bulkheads were quite strong, being stiffened and braced to the side. When connected with the transverse bulkheads, the entire system contributed materially to the overall strength of the ship. Indeed, on 26 August 1904 representatives from Cunard, John Brown, and Swan, Hunter all met with Mr Cornish, Chief Surveyor of Lloyd's Register in London; they desired to arrange 'a reduction in the original scantlings' of the sisters based on the strength of the longitudinal bulkheads being introduced to the design. A 'reasonable reduction' was eventually allowed based on both the longitudinal bulkheads and the use of high-tensile steel in the shell plating described earlier.[16]

It was decided to utilise these new wing compartments lining the boiler rooms as coal bunkers. This decision was not a problem for the Admiralty or, for that matter, for Cunard: coal sitting in the bunkers would help shield machinery from enemy fire. At the same time, other Cunard liners had sported longitudinal coal bunkers, and it was convenient since the coal was always close to the boiler furnaces. Each coal bunker was further subdivided into two semi-autonomous compartments with a partial transverse bulkhead.

Through this arrangement, *Lusitania* was endowed with thirty-four primary watertight compartments. When all of the subdivisions were accounted for, however, there were a total of 175 watertight compartments. Watertight decks at the E and F Deck levels also topped off most of the compartments. The Forepeak was capped off by E Deck; Cargo Holds Nos. 1 and 2 were capped by F Deck. Astern, aft of the Turbine Engine Room, F Deck was again made watertight. There was also a watertight deck above the outboard coal bunkers.

This brings us to a consideration of *Lusitania*'s safety, stability, and comparisons to other ships of the era as well as to modern-day Safety of Life At Sea (SOLAS) regulations. When she entered service in 1907, *Lusitania* was hailed by *The New York Times* as being 'as unsinkable as a ship can be …' A pre-maiden voyage Cunard booklet likewise said: 'In all, the *Lusitania* will have 175 watertight compartments, so that it may be claimed for her that she is as unsinkable as a ship can be.' Indeed, *Mauretania*'s watertight subdivision and basic structure was quite similar to *Lusitania*'s, and she

16 Executive Committee Minutes 1904–B4–22, Cunard Archives, University of Liverpool Library.

sailed the rugged North Atlantic from 1907 to 1934 – including time spent in war service – without ever suffering serious structural defects. This testifies to the strength and high quality of the ships' designs. However, *Mauretania* was also never involved in a serious incident at sea, such as being struck by a torpedo or mine, colliding with an iceberg, as *Titanic* did, or running into another large vessel. So how does *Lusitania*'s watertight subdivision stand up under close scrutiny?

The decision to adopt longitudinal coal bunkers was not unprecedented, either in the Royal Navy or on other liners. However, there were also risks. In order to pass the coal from the bunkers into the stokeholds, it was necessary to cut apertures into the base of these bulkheads, and through these openings it was possible for a flooded coal bunker to allow the sea into the inner compartment. Thus, these apertures were each fitted with a watertight door that could be sealed in the event of an emergency.

It was, however, a matter of historic fact that there could be problems in actually sealing these doors during a crisis. Let us review, briefly, the fate of the Cunard liner *Oregon*. On Saturday, 6 March 1886, she departed Liverpool with between 846 and 852 passengers and crew aboard.[17] In the early morning hours of 14 March, 18 miles off the coast of Long Island, a schooner came out of the darkness like an apparition, right off *Oregon*'s bow. The Cunarder's crew attempted to take evasive action, but *Oregon* was struck. The schooner sank promptly and without survivors, having disappeared immediately after the collision; it has been tentatively identified as the *Charles R. Morse*, which was also reported missing that night.

All attempts to stem the inflow of water through the three holes in *Oregon*'s iron hull failed. An attempt was made by Captain Cottier to beach the ship on nearby Fire Island, but just then water entered the undamaged Boiler Room and put the powerplant out of commission. The ship was evacuated and sank about four hours after the collision.[18] Not a single life was lost in the incident. However, the reason water was able to enter the Boiler Room from the damaged outboard areas had to do with the failure of the watertight doors to properly seat themselves. This was partly due to the distortion of the bulkheads in the collision, and partly to the build-up of coal dust and debris that clogged the doors' sills and tracks.

This experience demonstrates how vitally important it was to ensure that *Lusitania*'s watertight doors – particularly those that led from her wing coal bunkers into her Boiler Rooms – could be sealed effectively, reliably and quickly. Her bulkheads were fitted with sixty-nine watertight doors. The primary thirty-five doors were operated by the Stone-Lloyd system. The system operated on hydraulic power (not steam, electricity or compressed air), the pressure of which was constantly maintained by two steam-driven duplex pumps located in the Engine Room. From there, a pressure main that circled the ship led to a powerful hydraulic cylinder attached to each door; the closing of each door was controlled by an operating pedestal. This lever allowed pressure into the closing main, reversing the control valves and admitting pressure to the cylinders. When the system was activated, either from the local lever or from the Bridge controls, a continuous warning bell sounded to warn crewmen in the area to get clear. Once the doors were closed, anyone shut in the compartments could still escape by using the control lever to raise the door again while they exited; it would automatically close behind them as soon as they let go of the lever. The remaining thirty-four watertight doors were of the ordinary sliding pattern, worked by control rods and gearing from the Shelter Deck (C). A dial at the control point showed the position of the door at any given time. As specified by the British Board of Trade, all of the Stone-Lloyd doors could also be operated this way as a back-up.

Notably, only the coal bunker doors in Boiler Room No. 3 were of the Stone-Lloyd system; those associated with Boiler Rooms Nos. 1, 2 and 4 were of the older control rod type, a detail that may not have seemed important at the time. However, fitting only one quarter of what were arguably the most critical watertight bulkheads in the ship with the hydraulic system could hardly be described as wise. Additionally, in hindsight – knowing that *Lusitania*'s Boiler Rooms Nos. 1 and 2 and their starboard coal bunkers were the ones that bore the brunt of the torpedo's detonation – this omission seems particularly poor.

Assuming that all the liner's watertight doors were closed in the event of an emergency, *Lusitania* was designed to float with any two major watertight compartments flooded – even her largest ones. However, there were also flaws in the nature of her subdivision. Along

17 A closer breakdown of these passengers and crew shows: 186 First Class, 66 Second Class and 389 (or 395 according to some accounts) Third Class passengers. There were also 205 officers and crew aboard.

18 According to some accounts, she sank up to eight hours after the collision.

Cunard's resident Naval Architect, Leonard Peskett. When once asked if he had designed *Lusitania* and *Mauretania*, he replied: 'I had a great deal to do with it. The design was spread over a large number, but I was responsible.' (Authors' Collection)

with the introduction of the longitudinal compartments came the possibility of asymmetrical flooding, which is extremely dangerous as it causes a list to the wounded side as water collects there; this list would be exacerbated if water came in through open or damaged ports and collected above watertight decks, adding more top weight on that side. This process could eventually cause the ship to capsize. Realising this potential issue, the ship's designers made arrangements so that, through the use of pumps, the various watertight compartments in the ship's double bottom could be flooded. If *Lusitania* took a heavy list to one side, counter-flooding compartments on the other side might in theory trim the list.

After the *Titanic* disaster, the practicability of various forms of watertight subdivisions and lifesaving appliances came into question. In particular, questions arose as to whether longitudinal bulkheads would have saved *Titanic* from her fate. Harland & Wolff's Edward Wilding – who had helped design the *Olympic*-class ships – testified at the British Inquiry into the sinking. Cunard's resident Naval Architect, Leonard Peskett – the team leader on the designs for *Lusitania* and *Mauretania* – was also called to testify as an expert on the subject. Peskett was closely questioned regarding various aspects of *Lusitania*'s safety. Coming some three years before *Lusitania* sank, but after *Titanic*'s famous loss, his testimony is quite revealing. After Wilding testified that longitudinal subdivisions could cause a list, Peskett pointed out that a list could 'be counteracted':

> 21087. In what way? – Supposing that any two compartments were flooded on either side, the opposite side could be filled in several ways; also you could flood the engine room and the compartment immediately abaft it; also the shaft tunnel aft.
> 21088. (The Commissioner.) But that is rather heroic, is it not? – There would be less trouble in doing that latter part than any of the other parts.
> 21089. (Mr. Raymond Asquith.) Have you made any experiments to see how far that is practicable? – No, no experiments have been made.[19]

Even as late as 1912, then, 'no experiments' had been made to demonstrate the practicability of *Lusitania*'s counter-flooding system. Meanwhile, Edward Wilding cut straight to the heart of the matter:

> 20877. Did your firm build the '*Teutonic*' and the '*Majestic*'? – Yes.
> 20878. Were longitudinal bulkheads placed in those ships? – I think so, and they were promptly taken out, or holes made in them, at a later date.
> 20879. Can you tell us why they were dispensed with? – Because they were considered dangerous.
> 20880. Can you tell us why they were considered dangerous – from what point of view? – From the point of view that if one side of the ship got flooded it might lead to a very serious list, and possibly to a capsize.[20]

An enormous debate raged within the general maritime community over which form of watertight subdivision was superior. Following the conclusion of the British Inquiry into *Titanic*'s loss, *The Shipbuilder* magazine reported:

> The disadvantage of the longitudinal system is that, with several compartments flooded on one side only, the vessel will take an unpleasant list, and if she

19 British Inquiry into *Titanic*'s sinking.
20 *Ibid.*

> does not possess sufficient stability will completely capsize. It is our own opinion that the longitudinal arrangement is the better one, provided it is adopted in conjunction with a watertight lower deck and that the ship has a suitable metacentric height … It seems to us … that [counter-flooding] arrangements would in most cases be ineffective, as owing to the list of the ship practically no water would run in on the undamaged side and the compartments on that side could only be filled by pumping up.
>
> Another danger with longitudinal bulkheads is that, as the side spaces are used as coal bunkers, the watertight doors through which the coal passes to the stokeholds will nearly always be open, and it might be difficult to shut them owing to the presence of coal in the openings. It is claimed that this difficulty has been largely overcome, and can be further minimized, by the provision of screens inside the bunkers placed above the doors, an arrangement always adopted in warships, but some doubt must always exist on this point. It would, therefore, be an advantage in this respect if oil were adopted instead of coal as fuel, in which case the doors could always be kept closed or dispensed with entirely.

Again the thinking on the matter at the time of writing is highly revealing. Even before *Lusitania* was sunk, it was a known fact that the watertight subdivision of her and *Mauretania* was potentially hazardous, because the sisters' metacentric height in an undamaged condition was only some 2ft. *The Shipbuilder* had commented on the metacentric height of *Lusitania* and *Mauretania*:

> To gain steadiness at sea, it is usual to design passenger steamers with fairly small metacentric heights; but there are two objections to this course. Firstly, such vessels when exposed to a beam wind often take an unpleasant list on account of their large exposed superstructure; and, secondly, if the ship gets damage the loose water in the compartment in communication with the sea may reduce the stability to a dangerous extent. Neither of these two defects need be feared in the *Mauretania* [or *Lusitania*], as the metacentric height is sufficient under all working conditions to prevent the ship heeling over under the action of a beam wind, while the great inertia due to her enormous weight and dimensions gives her a period of rolling equal to that of the steadiest liners afloat …
>
> The stability when damaged has also been carefully investigated; and although the vessel may take an unpleasant list should two of the wing compartments be flooded when the bunkers are empty, she will always have a good positive metacentric height with a long range of stability.

This text put an absurdly positive spin on the original calculations made by Leonard Peskett and his team. At the time the designs for the ships were finalised, a series of stability analyses were performed to ascertain how they would react under various potential damage scenarios. These calculations demonstrated that if one coal bunker was flooded, the ships would be pulled over some 7 degrees away from the vertical. With two flooded, that list would increase to 15 degrees. With three coal bunkers flooded, the analysis showed the ships would become dangerously unstable. In fact, Peskett advised that if they assumed a list of more than 22 degrees and held that position, they should be abandoned without delay.

A modern stability analysis of *Lusitania* confirmed Peskett's original fears.[21] With one coal bunker flooded in arrival (light) condition, the calculations showed the ship's metacentric height was 1.0ft; with two flooded, that margin fell to 0.35ft. With the third bunker flooded, that figure turned negative, moving to -0.21ft. This meant that the ship would have become highly unstable, unless quick action was taken through the intervention of the 'heroic' counter-flooding measures. However, counter-flooding measures depended on 1) the presence of crew members to implement the orders, assuming that they had not already abandoned their posts; 2) the ability to effectively communicate orders throughout the ship; and 3) the operation of the ship's primary systems in order to provide motive power to the pumping system. It also depended on actually being able to close and seal all of the watertight doors, including those not operated by the hydraulic system. All of this, in retrospect, seems like something of a recipe for disaster.

Remarkably, the comparison between the two types of watertight subdivision – transverse only as opposed to transverse and longitudinal with coal bunkers in the outer compartments – left the realm of hypothetical and became

21 *The Saga of the RMS* Lusitania, *A Marine Forensic Analysis* (William H. Garzke Jr, Robert O. Dulin Jr, Peter K. Hsu, Blake Powell, F. Gregg Bemis), 8 January 1998.

a solid reality when the liner *Justicia* was torpedoed during the First World War. *Justicia* was built by Harland & Wolff, of Belfast, Ireland. Laid down in early 1912 on *Titanic*'s former slip, *Justicia* was originally ordered by the Holland-America Line as *Statendam* II. Her hull number, 436, was only three digits removed from the last of the *Olympic*-class liners, *Britannic*. Her relationship to the three *Olympic*-class vessels was undeniable.

Justicia was 776ft in overall length, with a beam of 86ft and a gross tonnage of 32,234. Being designed and built by Harland & Wolff to the same general lines as the *Olympic, Titanic* and *Britannic*, she was comparable in size to *Lusitania* (the Cunarder was 11ft longer, 1½ft wider, and 684 tons smaller). She thus, for the purposes of our discussion, represents an 'alternative version' of *Lusitania*. The intended Holland-America liner was not built to Admiralty specifications, and thus had transverse longitudinal bunkers, without flanking wing bunkers; her configuration was more like that adopted by *Olympic* and *Titanic*.

Statendam was launched on 9 July 1914, but soon after the outbreak of war, the British Government requisitioned the unfinished liner for wartime service. *Statendam* was renamed *Justicia* and handed over to Cunard's control – ironically enough, as recompense for the loss of *Lusitania* – to serve as a troopship. However, Cunard could not put a crew together quickly enough for the task, and *Justicia* was, instead, handed over to White Star Line's control. Many of those assigned to *Justicia* were formerly crew members aboard the lost *Britannic*.

Justicia was completed in April 1917 and entered service in the dingy grey guise of a troopship. In early 1918, she was 'dazzle'-painted to confuse submariners who might have been watching her through their periscopes, but the technique did not succeed. On 19 July 1918, while westbound as part of a convoy, and under the command of Captain Hugh David, the German submarine *UB-64* evaded the convoy's destroyer screen and fired two torpedoes at her.

The first torpedo struck the ship on the port side at 2.30 p.m. The Engine Room, Dynamo Room and Cargo Hold No. 4 were all flooded, killing ten men. However, *Justicia* remained afloat with her watertight doors closed. The lookouts spotted a second torpedo as it was incoming and the deck gun crews fired a precision shot that detonated the torpedo before it hit the ship. Destroyers then dropped a barrage of depth charges and began escorting the listing *Justicia* to safer waters.

Two hours later, at 4.30 p.m., the same submarine, commanded by Kapitänleutnant Otto von Schrader, evaded detection and attacked the ship again, firing two more torpedoes into her side. Again, the ship refused to sink, although, with her engines disabled, the vessel was evacuated, except for a skeleton crew. She was taken under tow by HMS *Sonia,* with the intention of grounding the ship in shallower waters of Lough Swilly.

At 8 p.m., von Schrader and *UB-64* attacked a third time, firing a fourth torpedo into *Justicia*. Still, the liner refused to sink. This time, depth charges from *Justicia*'s escorts damaged her attacker. *UB-64* abandoned her five-and-a-half-hour molestation of the liner and withdrew to a safe distance, where they could report the ship's position to other U-boats in the vicinity. The tow project resumed.

At 4.30 a.m. the following day – some fourteen hours after the initial torpedo struck – a second submarine, *UB-124*, attacked. Its torpedo was set to run too shallow, however, and missed the target completely. By 6 a.m., twelve destroyers and fourteen other vessels surrounded *Justicia*. The tow process continued, but was now hampered by poor weather. Finally, at 9.15 a.m., the determined *UB-124* attacked again, firing two torpedoes. At the same time, a third U-boat, *U-54,* also fired two torpedoes at *Justicia*. While *UB-124* is typically credited with the kill, contemporary accounts and documentation also credit *U-54*. The truth is still disputed.

In any event, two of these four fresh torpedoes struck *Justicia*. She now began to settle heavily by the stern and also to port, and the remaining crew members were forced to abandon ship. At 10.30 a.m., Captain David himself was evacuated; the tow process continued, however, in the hope that she could still reach shore. At 12.40 p.m., 20 July 1918, *Justicia* finally succumbed to her wounds and sank.

The only lives lost were in the initial explosion. In all, six torpedoes had struck, yet from the time she took the first torpedo on 19 July to the time that she sank on 20 July, some twenty-two hours and ten minutes had elapsed. *Lusitania*, as is so famously recorded for us through the pages of history, took one torpedo and immediately heeled over because two of her wing coal bunkers flooded, critical watertight doors were apparently not closed and counter-flooding was impossible. She sank in just eighteen minutes, taking the lives of nearly 1,200 passengers and crew.[22]

22 *Into the Danger Zone: The Lusitania, First Battle of the Atlantic, and Liners During the Great War*, Tad Fitch and Mike Poirier, Second Edition (Blurb, 2023). The details of *Justicia*'s sinking are drawn from this volume.

Clearly, *Lusitania*'s design under Admiralty supervision was focused on protecting interior compartments from enemy shellfire. As submarine warfare was not really an essential issue before the outbreak of the First World War, little or no investigation was done into the potential damage that a submarine-launched torpedo could cause to the ship. Under this rather unanticipated scenario, *Lusitania*'s design was highly vulnerable.

Interestingly, *Aquitania* was designed along very similar lines to *Lusitania* and *Mauretania,* albeit on the scale of *Olympic* and *Titanic*. When she was built, *Titanic*'s loss was fresh in mind, and her design reflected the new thinking on watertight subdivision. Leonard Peskett did an extensive analysis of her stability under certain damage scenarios. The most extreme scenario considered was the flooding of not one, two or three wing compartments, as had been considered with *Lusitania* and *Mauretania*, but *all* of them. With some 5,300 tons of water filling these spaces, *Aquitania* would still only take on a list of 26 degrees away from the vertical. This list would certainly be severe, but it was calculated that even under this 'almost impossible' scenario, as *Engineering* magazine termed it, her overall stability would not be in question as it was with only three wing compartments flooded aboard the earlier ships.

Lusitania and *Mauretania* would not meet modern SOLAS regulations on watertight subdivision. In one of the most remarkable twists of irony, *Titanic*'s watertight subdivision would pass, or nearly pass, modern regulations. As *The Shipbuilder* stated in the previously cited article on watertight subdivision, the inner skin should be far enough away from the exterior hull ('10 to 15ft' or some 20 per cent of the vessel's overall beam as specified by SOLAS in modern commercial ship design) that damage to the outer shell would not damage the inner shell. Specifically in Boiler Room No. 1, this was a problem in *Lusitania*'s design.

As *Lusitania*'s bow narrowed in that region, it was found necessary to 'step in' the longitudinal bulkheads on either side of the ship at a point roughly 40ft forward of the compartment's aft transverse bulkhead. Because the outer hull was narrowing and the inner bulkheads were parallel to the ship's longitudinal axis, there were spaces in Boiler Room No. 1 where the bulkheads were dangerously close to the ship's outer hull. If a torpedo detonated in this region, it was quite probable that the inner bulkhead would be severely compromised by the initial explosion, and that water would be allowed into the inner compartment either through damaged watertight doors or because the inner bulkhead itself had been compromised.

In summary, *Lusitania*'s system of watertight subdivision was not as good as was commonly believed at the time of her construction. Her design left her vulnerable to any flooding, and particularly to a torpedo strike, anywhere along the length of her boiler and engine compartments. She was quite vulnerable to asymmetrical flooding, and without proper planning to ensure a practical method of counter-flooding.

As to *Lusitania*'s stability and metacentric height, many allegations of poor design have been levelled against her designers, specifically Leonard Peskett, labelling the ship dangerously top heavy. However, this was untrue under normal service conditions, undamaged; *Mauretania*'s long service career, and *Lusitania*'s own service between 1907 and 1915, proved that point. *Lusitania* was also quite unthreatened by flooding of her extreme forward or aft compartments, the ones not lined by longitudinal bunkers. Her design was even enough to withstand asymmetrical flooding of two wing bunkers at any given time, as long as there were no other contributing factors to consider. However, with any additional damage or a third compartment flooded, *Lusitania*'s stability immediately became critically suspect, and this risk was known before she and *Mauretania* were ever built. All of *Lusitania*'s remarkable strength, her high-tensile doubling steel plating and her reinforced decks were not enough to save her from the one critical emergency situation that she actually faced during her career.

Astonishingly, Cunard later nearly lost *Mauretania*, as well. In November 1916, while sailing for Halifax, Staff Chief Engineer Andrew Cockburn, a recent survivor of *Lusitania*'s sinking, found water entering the ship while making his evening rounds. Apparently, when the coaling at Liverpool had finished, an errant worker had not properly sealed a coaling hatch, and no one had caught the mistake. Although the seepage was slow, the situation worsened overnight: the pumps became clogged with ash and coal debris and stopped working, and the water grew so deep in the Boiler Room that the lower level of furnaces in the boilers were extinguished. The ship took on a worrisome list, and Captain James Charles was forced to slow *Mauretania* and turn her so that the wind was pushing against her low side. After working heroically through the night, the engineers managed to get the pumps running again, got ahead of the flooding, and the ship made Halifax successfully. But the low margin of *Mauretania*'s metacentric height, when combined with failed pumps, had contributed to a close shave.

In July 1921, after a fire broke out aboard *Mauretania* while she was berthed in Southampton, the ship was again nearly lost. For six hours, firefighters doused both the flames and woodwork in areas above the fire with water. This water collected on D and E Decks; it would ordinarily have drained through the watertight decks via scuppers into the bilges, but the scuppers became clogged with debris. Two mooring lines snapped, and the liner took on a list of about 15 degrees to starboard. All the water suddenly rushed to that side, collecting to a depth of 6ft. Only the heroic efforts of one of her officers and two crewmen to open a gangway door allowed the water to drain and saved the ship. Again, the combination of *Mauretania*'s low margin of metacentric height, her watertight decks and clogged scuppers had nearly caused her to roll over and sink at her pier.

THE HEART OF RMS *LUSITANIA*

Lusitania's revolutionary powerplant – her very heart and life – quite deservedly stole the show when she entered service.

Charles Parsons, inventor of the marine steam turbine. (Authors' Collection)

Right through the 1890s, all the first-rate express ocean liners had been double-screw vessels propelled by reciprocating engines. However, it was becoming clear as the new century came on the scene that the development of the reciprocating engine was reaching its zenith. There was only so much horsepower that such engines could produce. They also had disadvantages, including vibration at high speeds and the fact that they took up enormous space within a ship's hull.

During *Lusitania*'s design phase, the question of how much horsepower would be needed to meet her required speed came to the fore. Would reciprocating engines suffice? When it was believed that the liner's width would be some 78 or 80ft, calculations showed that 60,000hp would be required. To meet that demand, three quadruple-expansion, five-cylinder reciprocating engines could be used, each tied to its own propeller, creating a triple-screw design.

When it was discovered that the ships needed to be 88ft wide to create a more stable vessel, the equation changed entirely. All new model tests and computations were carried out. Swan, Hunter – *Mauretania*'s builders – did a rather unique set of model tests and found that when travelling at 25 knots against a wind of 25 knots, she would require 12 per cent more power than without the headwind. When travelling at 25 knots with a following wind of the same speed, she would need only 4 per cent less power. These results would also very closely approximate *Lusitania*'s requirements. The tests showed that *Lusitania* and *Mauretania* would need some 68,000 nominal horsepower to achieve the contractual speeds. This was a problem since the triple-screw design producing 60,000hp was already pushing the limits of reciprocating engine technology. The only way to obtain the required horsepower was to add more engines; yet a quartet of five-cylinder reciprocating engines, along with the boilers to fire them, would have consumed too much interior space to be financially viable.[23]

It turned out that there was another possible alternative, although it had never been tested on this sort of scale before. This new possibility was the marine steam turbine, pioneered by Sir Charles Algernon Parsons with the 100ft test craft *Turbinia* at the 1897 Naval Review. Since then, other small- to medium-sized vessels had begun to experiment with this technology, and with good results.

23 By way of comparison, *Kaiser Wilhelm II* (1903) sported only 40,000hp; the *Olympic*-class ships, launched starting in late 1910, had two four-cylinder, triple-expansion, reciprocating engines each, with each engine supplying a nominal yield of 15,000hp.

Although the idea looked as if it could work on paper, the question was: would it actually work on this scale? If *Lusitania* or her sister did not meet the speed requirements outlined by Parliament, there could be severe monetary penalties or the ships could be rejected. Then, too, there was the question of reliability, as there was not much practical data to go on yet.

In 1903, a committee was set up to investigate the propulsion matter thoroughly, and to make a recommendation to the Cunard Line. This committee was comprised of specialists in the field of ship design and propulsion, and included James Bain, Marine Superintendent of the Cunard Company; Engineer Rear-Admiral H.J. Oram, CB, Deputy-Engineer-in-Chief to the British Navy; J.T. Milton, Chief Engineer-Surveyor of Lloyds of London; H.J. Brock of Messrs. Denny, of Dumbarton, which company had experience building ships fitted with turbine engines under licence from Charles Parsons. A representative of John Brown, T. Bell; one from Swan, Hunter, Sir W.H. White, KCB, and Andrew Laing of the Wallsend Engineering Company, were also included. Lastly, and perhaps most importantly, Sir Charles Parsons himself was brought in to consult the subject.

After lengthy deliberations, the committee decided that Parsons-style steam turbines would be adopted in both new speed queens. Even so, many questions remained. They already had data comparing the two types of engines in small vessels such as the cross-Channel ferries *Arundel* (1900, reciprocating engines) and *Brighton* (1903, turbine engines); data about the use of turbines in ocean-going steamships only started to become available when the Allan liner *Victorian* made her maiden voyage in April 1905, however. She was the first transatlantic vessel to use turbines, and her sister ship *Virginian* quickly followed. Cunard also experimented with two nearly identical sister ships, *Caronia* and *Carmania*, fitting the first with reciprocating engines in a double-screw configuration, and the second with turbine engines in a triple-screw configuration. From these ships, lessons could be learned quickly and applied to *Lusitania* and *Mauretania*, even though the decision to equip the ships with turbines had long since been settled.

Caronia made her maiden voyage on 25 February 1905. *Carmania* made her maiden voyage on 2 December of that year. Once the two ships had entered service, it was quite clear that turbines were more than capable of providing high speed. *Carmania* proved generally faster than her more traditional sister. At high speeds, her turbines proved more efficient than the reciprocating engine; however, it was found that at lower speeds the tried-and-true reciprocating engine still had the cost advantage.

In the meanwhile, work on building *Lusitania* and *Mauretania* had initially been focused on their bow and midships sections. *The Shipbuilder* reported that even after the final decision in favour of turbines, and even after the orders had been placed for both ships:

> many details in regard to the propelling machinery were … still in abeyance, and this explains why the photographs taken in the preliminary stages of the construction show a considerable advance in the progress of the work at the forward end … At the after end the work was somewhat delayed, in order that the position of bulkheads, engine seatings, etc., could be fixed to suit the best arrangements for the machinery which had not then been finally settled.

Although specifically discussing *Mauretania*, the statement applied equally to both ships. While certain details still needed to be worked out as construction began, the major decisions were already solidly in place.

In the end, *Lusitania*'s powerplant was quite different from *Carmania*'s. Four turbines would be installed in the larger vessel for forward thrust; the power would be transmitted to the sea through four propellers, each linked to its own forward-thrust turbine. *Lusitania*'s quadruple-screw arrangement, which *Mauretania* also adopted, was unprecedented. Previously, almost all crack Atlantic liners had been limited to two screws, and even *Carmania* had only three. The radical decision to use four screws was prompted by necessity: to divide 68,000hp four ways meant that each propeller shaft would need to transmit only about 16,000–17,000hp. If fewer shafts had been employed, the torque stresses imposed on them would have been beyond the tolerable limits of the steel used. If a shaft had snapped or otherwise failed at sea, the results could have been disastrous: more than one ship has gone to the bottom after a broken propeller shaft allowed the sea to flood the Engine Room.

After careful study, it was determined that two high-pressure turbines would drive the outboard set of propellers, while two low-pressure turbines would drive the inboard set. One drawback of adopting turbine engines was that they were irreversible for astern thrust, so two high-pressure reversing turbines were installed just forward of the low-pressure ahead turbines that worked the inner propellers. Thus a total of six turbines were installed. By adopting four units for forward thrust, along with four screws, the propelling machinery could be divided into two complete sets, one on either side of the ship. With the Engine Room subdivided into smaller compartments, even if damage was taken to one side of the powerplant, the other side would most likely be able to function.

The placement of the propellers was important in achieving the ship's desired speed. It was decided that the outer propellers would sit some 70ft forward of the inner propellers.[24] The shafting for the inner propellers was entirely contained within the ship, and their framing was bossed out and supported by web frames. The forward propellers were also carried by heavy webs. Each propeller was given three bolted-on, manganese-bronze blades, which were shaped quite roundly.[25] As originally fitted, the outer propellers had a diameter of 15ft 0in and a pitch of 16ft 6in, while the inner propellers had a diameter of 16ft 6in and a pitch of 15ft 9in.[26] The blades of all four propellers were contained within the beam-line of the ship, but all worked within free water. After extensive model testing, it was decided to rotate the outer propellers in an inward direction, while at the same time having the inner propellers rotate outward, or in the opposite direction; initially it was thought that this would achieve optimum efficiency. In 1913, however, the outer props were changed so that they, too, spun in an outward direction just like the inner props.[27]

In order to determine the final dimensions of the turbines, it was necessary to decide how fast the propellers would spin to efficiently transmit the horsepower into the sea. At the time, space constraints and the need to adopt small turbines led Channel steamers and small craft to adopt a high speed of rotation. There were those who felt that high rotation speed meant losing a percentage of overall efficiency. However, after study, it was decided to adopt a propeller rotation speed of about 185rpm on *Lusitania*. Working backward from this, the final dimensions of the ship's turbines could be calculated.

The turbines' rotor-drums were manufactured at John Brown's Atlas Works in Sheffield through hydraulic forging; the high-pressure drums had a diameter of 96in. The larger low-pressure turbines had rotor-drums of 140⅜in. These were all hollow forged, and the rotor-drums for the low-pressure ahead turbines were the largest yet constructed. The Atlas Works also made all of *Lusitania*'s rotor spindles and propeller shafting.

The blades for the turbines ran from 2¼in up to 22in long. For the longer blades, three rows of shrouding, consisting of circumferential strips laced with copper wire that was then soldered, was employed to give them both radial and lateral rigidity. To prevent distortion from the expansion of the drum and of the brass strip, expansion joints were included in these bindings. The strip was divided into shorter lengths, which were connected by brass tubes, within which they could slide. Great care was taken during the construction of the turbines to ensure balanced operation.[28]

For times when the turbines needed maintenance, special lifting gear was designed to raise the outer casings for the turbines by a height of 4–6ft (depending on which turbine was under repair); the rotors themselves could also be lifted if necessary, once the bearings had been removed. This could even be accomplished, if it was deemed necessary, at sea, and did happen on one occasion during her actual time in service.

After this consideration, one might wonder: just how closely did the turbines' performance match their anticipated performance? During *Lusitania*'s trials, a torsionmeter calculated the amount of power being developed; this particular information was recorded beginning on 27 July 1907, when the ship had a draught of 32ft 9in and a displacement of 37,080 tons. This information was also compared to the ship's speed at the time of reading, the revolutions per minute and the rate of the propellers' slip through the seawater. The results are astounding:

24 The placement of *Mauretania*'s propellers was somewhat different: the centres of her inner props sat 12ft 10in forward of the after perpendicular, while the centres of the outer props sat 78ft 11in forward of the inner props, or 91ft 9in forward of the after perpendicular.

25 *Mauretania* initially had similar propellers, except that the blades were slightly narrower and longer than those on *Lusitania*. Both ships' propellers were replaced numerous times during their careers to improve performance, as it was found that the originals were far from ideal.

26 In another source, a slightly different set of data is given, mainly a diameter of 16ft in for both sets, and a mean pitch for the inner propellers of 14ft 6in, while for the outer props the mean pitch was 15ft 9in. A root thickness of 9.75in and a tip thickness of 0.75in were also specified in this document.

27 See further details on this in the section covering the 1913 turbine repairs.

28 A different approach was taken to *Mauretania*'s turbines. Her discs, gudgeons, shafts and drums were constructed of Whitworth fluid-pressed steel. To achieve the maximum strength with as little weight as possible, it was decided to reduce the number of parts to the bare minimum.

Time	RPM	Speed in knots	Shaft horsepower	Slip of propellers
1st double run	194.3	25.62	76,000	17.2%
2nd double run	186.0	25.0	65,500	15.5%
3rd double run	174.2	23.7	51,300	14.5%
4th double run	161.5	22.02	40,500	14.3%
5th double run	147.6	20.4	29,500	13.1%
6th double run	131.1	18.0	20,500	13.7%
7th double run	116.1	15.77	13,400	14.6%

This information is even more remarkable when one considers the relative newness of the technology and the unprecedented proportions of the turbine engines powering *Lusitania*. The builders were really in uncharted waters, and not only did they correctly design a powerplant that developed more than the needed horsepower, but through the efficient design of the ship's hull, she was able to exceed her mandated service speed. In fact, it was calculated after the ship's trials that in order to maintain a speed of 25.4 knots, *Lusitania* would need to develop some 68,850hp; this is incredibly close to the original calculations based on the model tests that to maintain 25 knots she would need some 68,000hp.

In order to motivate the turbines, vast quantities of steam at high pressure were required. To provide this steam, twenty-three double-ended and two single-ended coal-fired boilers were provided. These twenty-five boilers were arranged in four separate watertight compartments, with Boiler Room No. 1 being forward and No. 4 being aft, closest to the Engine Room. Six boilers were arranged in each of the aft three Boiler Rooms (Nos. 2–4), in two rows of three abreast. Because of the taper of the ship's hull in the region of Boiler Room No. 1, a different arrangement was adopted there. In the stern portion of the compartment sat a single row of three double-ended boilers. Forward of these, where the longitudinal bulkheads stepped in, a second row of two double-ended boilers was installed. To provide the last fraction of steam required to develop the needed horsepower, two single-ended boilers were placed adjacent to the forward transverse bulkhead of the compartment.

The double-ended boilers were 17½ft in diameter and 22ft long. The single-ended boilers were identical, except that they were slightly shorter at 12ft. The total 192 furnaces had a collective grate area of 4,048 sq. ft, while the total heating surface was 158,350 sq. ft. The boilers' shells were made of high-tensile steel, which had a maximum tensile strength of 36 tons psi.

Once the water within the boilers had been superheated into high-pressure steam, it was carried aft via a network of steam pipes that ran above the tops of the boilers, roughly 10ft below the load waterline of the ship. Each Boiler Room had its own steam pipe. On the forward side of the forward Engine Room bulkhead, a stop valve was fitted to each line. These valves could be worked both from the Engine Room and from the Boat Deck. Each Boiler Room was tied into its own funnel to ventilate the coal smoke and fumes from the combustion chamber, and the funnels' height above the grate level was some 130ft. With the powerplant operating under full steam, it was rated to deliver steam at a pressure of 195psi to the turbines.

Once the steam had passed through the turbines, it was sent to the condensers to be turned back into feed water for the boilers. The condensers were arranged in pairs, and gave a total of 82,800 sq. ft of cooling surface. There were also two auxiliary condensers placed in the forward end of the Engine Room. These auxiliary units had a cooling surface of 4,000 sq. ft, and were given their own separate circulating pumps and air pumps.

It was estimated that *Lusitania* would burn about 1,000 tons of coal each day of her crossing, and enough bunker space was devised to give the ship a carrying capacity of some 6,000 tons of coal. Further arrangements were made to carry extra coal in the No. 2 hold and above Boiler Room No. 2, if the need arose for additional capacity. In actual service, *Lusitania* sometimes burned coal less eagerly than anticipated, even dropping at times into the mid-800 tons per day range; at other times she used the full amount that she was expected to burn. The ship's Engine Room Log for her third westbound crossing, in November 1907, is telling. From Liverpool to Queenstown, Ireland, she consumed 408 tons of coal; between Queenstown and New York, she burned 4,976 tons. Her galleys and auxiliary machinery burned 18 tons during the voyage, for a total consumption between Liverpool and

A view inside one of *Lusitania*'s boiler rooms showing the furnace doors. In the distance some of the watertight doors giving access from the stokehold to the wing coal bunker through the longitudinal bulkhead are clearly visible. (Authors' Collection)

New York of 5,402 tons. During each of the four full days of her crossing, she burned exactly 1,090 tons of coal, and her average speed for the entire crossing from Queenstown to New York was 24.25 knots, beating her previous average of 24.002 knots during her second westbound crossing when she had retaken the Blue Riband from Germany.

After the First World War, *Mauretania*, like all of the other crack Atlantic liners, was converted to burn oil fuel. This economised the operation of her stokeholds, and improved her performance markedly. It would have been interesting to see how *Lusitania* would have adapted to a similar conversion, had she survived the war. She doubtless would have enjoyed a performance increase at least equal to that enjoyed by *Mauretania*, although whether or not she could have rivalled *Mauretania*'s speed records is debatable.

There has been a long-standing feeling that *Mauretania* was the faster of the two sister ships. A brief investigation of this point is in order during our consideration of *Lusitania*'s powerplant. Let us start by comparing *Lusitania* and *Mauretania*'s trials. Between midnight on 29 July and 1 a.m. on 1 August 1907, *Lusitania* managed to make an average speed of 25.4 knots over a period of roughly forty-eight hours, in conditions described as 'favourable, with cloudless days and starlight nights'. In early November of the same year, *Mauretania* managed to average 26.04 knots during a similar type of test, but in conditions described as a 'moderate gale of force 7' during early portions of the event. This might seem conclusive, but another test favoured *Lusitania*. On a shorter run of 59 miles made twice in each direction during her trials, *Lusitania* averaged 26.45 knots. On the measured mile, run twice in each direction, *Mauretania* averaged 26.17 knots, slightly slower than her older sister.

It must also be pointed out that trials are not always indicative of actual service records, since the crew is not always familiar with the machinery that they are operating, the layout of the ship, or even with each other. It is also important to point out that ships were not always supplied with a grade of coal that was equal in quality to the coal that they would use while in service. So what did the two ships prove during their actual service careers? *Lusitania* entered service first, and took the Blue Riband on her second westbound crossing with an average speed of 24.002 knots.[29] In July 1908, she bettered her own record again, this time at an average speed of 25.65 knots and a time of four days, sixteen hours and forty minutes. This average continued to improve as the ship became further broken in and her original propellers were replaced with more efficient ones.

Captain S.G.S. McNeil, who served as *Lusitania*'s Chief Officer from late 1907 until the spring of 1911 during fifty-three voyages, and who later served as *Mauretania*'s Staff Captain and Captain, felt that *Lusitania* was not a bit slower than *Mauretania*. He believed that this 'wrong impression was probably brought about by the fact that the *Lusitania*, on her first eight voyages, did not have the type of propeller blade that the *Mauretania* had. When the new ones were shipped the former's averages improved considerably, and I remember three of them were 25.89 knots [in average] with only a decimal difference.' However the numbers tell a different story.

Mauretania's two fastest eastbound pre-war crossings were made at 25.88 and 25.89 knots, respectively. The first of these two crossings, made on 16 June 1909, was completed in four days, seventeen hours and twenty-one minutes; the second, made on 4 August 1909, was made in one minute less time. Her fastest pre-war westbound crossings, although against the prevailing current, were even faster. Departing Liverpool on her twenty-sixth voyage on 25 September 1909, she made the crossing in four days, ten hours and fifty-one minutes at an average speed of 26.06 knots, taking the westbound Blue Riband from *Lusitania*. On 10 September 1910, another westbound crossing proved even faster at four days, ten hours and forty-one minutes at the same average speed of 26.06 knots. In May 1911, she made an average of 27.04 knots (or just a shade over 31 land miles per hour) for a single day's run.

While Captain McNeil clearly felt that *Lusitania* was not slower than *Mauretania*, the simple truth of the matter is that after September 1909, *Lusitania* was never able to take the Blue Riband back from her slightly younger sister – an honour that *Mauretania* held until June 1929. There was certainly plenty of opportunity for her to try after September 1909; indeed, she could have done so at any point up until the Cunard Line reduced her top speed to 21 knots in November 1914.

Only the ignorant would claim that *Mauretania* was significantly faster than *Lusitania* at the time that the two ships served together. However, upon a close inspection of the original records, it is very clear that *Mauretania* was the faster of the two ships, specifically during their concurrent pre-war service.[30] Even so, *Lusitania* was able to turn in remarkably consistent and swift speeds and was only marginally slower than *Mauretania*. In the end, *Lusitania*'s remarkable powerplant was not only a marked departure from the standard technology of the day, but it was also a striking success, becoming a trend-setter for all the Atlantic liners that followed in her wake for the next forty years.

LUSITANIA'S SECONDARY SYSTEMS

In addition to her primary systems, *Lusitania* had many other systems that, while secondary, were essential to her operation, and to the safety and comfort of her passengers.

THE ELECTRICAL SYSTEMS

Since the Cunarder *Servia* of 1881, crack ocean liners had utilised electrical lighting. Beyond convenience, this also improved safety since up to that time open-flame light systems had been used aboard ships. Over time, electricity became indispensable; demand for electricity continued

29 Some other sources cite 23.993 knots as the average speed for this voyage, but Cunard's official statement of 24.002 seems more reliable.

30 *Mauretania*'s average speed increased significantly in later years as improvements were made to her powerplant and propellers; these were improvements that *Lusitania* never saw because she was lost during the war. Pre-war, *Mauretania* was still faster, but only by a small margin.

One of the telephones located in the finest First Class staterooms. These were connected to a central exchange board to direct shipboard calls. Astonishingly, while in port, the system could be tied into a landline, allowing passengers to call friends or relatives ashore right from their cabin. (Authors' Collection)

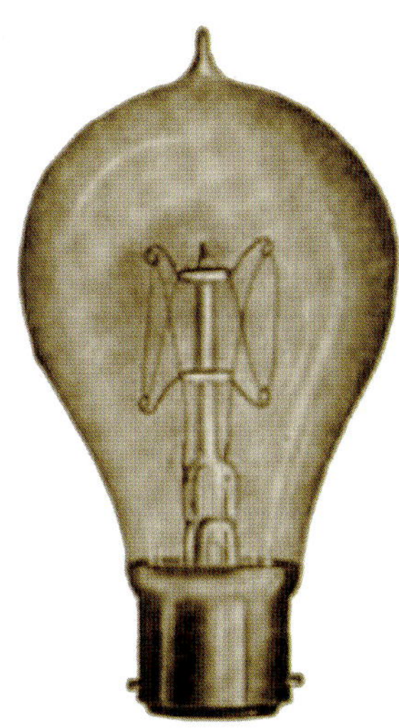

An advertising image for one of *Lusitania*'s 'tantalum' light bulbs. (Authors' Collection)

to grow as more and more uses were found for electrically operated devices and equipment.

Lusitania's electrical installation was unprecedented in power and scale, and comparisons were made at the time to land-based generating stations for a city of 100,000 residents. While Swan, Hunter decided to have *Mauretania*'s electrical installation fitted by an outside contractor, John Brown's Electrical Department undertook the complete installation on *Lusitania*.

A quartet of Parsons-style turbo-generators, rated at 375kW each, was placed in the Auxiliary Machinery Room, abaft the Turbine Engine Room. The compartment was divided along the centreline of the ship by a longitudinal watertight bulkhead, with two turbo-generators on either side. In this way, if flooding in one compartment knocked out one pair of turbo-generators, half of the electrical plant would remain intact.

Each turbo-generator was capable of supplying 4,000 amperes at 110 volts while turning at 1,200 rpm. The entire electrical-generating plant was operated by steam produced by the boilers. Unfortunately, it meant that if serious damage knocked out the ship's steam-generating plant, the electrical-generating plant would also fail; yet in all fairness that sort of event would have to be catastrophic.

From the turbo-generators, the power went to the primary switchboard, which was of the same design as those used by Royal Navy warships. One board was fitted in each of the Dynamo Rooms; the two boards could be coupled for tandem operation, or they could work independently – a very important consideration in case one of the compartments flooded or was otherwise put out of commission. Governing the connection between the two main boards was a double-pole 5,000-amp switch. Each board had a 4,000-amp main breaker and a dozen 1,000-amp subsidiary breakers. There were also twelve auxiliary switchboards distributed around the ship. Each of these subsidiaries had single-pole 150-amp switches as well as double-pole fuses of the Dixon blotter type. Electrical current was then carried to ancillary junction and distribution boards, then finally delivered to the instruments calling for current.

The main electrical cables were 2in in external diameter, insulated with vulcanised rubber with a double covering of asbestos and jute braiding thrown in for fireproofing. They were held in place by iron racks mounted on each side of the Engine Room and Boiler Rooms, and were 4 miles long. There were over 150 miles of electrical cables installed throughout the ship.

This system had to power an unprecedented number of sea-going devices. These included thousands of lamps for cabins and public spaces. Some 5,651 light bulbs were installed; of these were 4,635 16-candlepower 'tantalum' bulbs and 1,011 8-candlepower 'huntalite' bulbs. The 25-volt tantalum bulbs were made by Siemens Bros. Dynamo Works, Ltd, London. The filaments in these were unusually strong, and the bulbs were smaller than the typical carbon filament bulbs. There were also five 32-candlepower bulbs for the ship's navigation lights, including her red (port) and green (starboard) sidelights.[31] In the event that one of these lights burned out, a spare filament was turned on automatically, and an indicator on the Bridge immediately informed the ship's officers that the primary bulb needed to be replaced.

Lusitania was also graced with the rather novel feature of telephones for the use of both crew and passengers, separate from the Graham's loud-speaking telephones on the Bridge which have already been described. There was an exchange-less system of twenty-three telephones interconnecting the Bridge, Marconi Shack, Purser's Bureau, as well as the quarters of the senior officers and stewards. Additionally, there were eighty-nine telephones fitted in the Regal and *en suite* First Class staterooms, allowing passengers to communicate with other staterooms, as well as the Purser's Bureau, the Ship's Surgeon, Chief Steward, etc. To direct this flow of telephone traffic, there was a central exchange board. It was said of the First Class cabin telephones:

> The induction coil, condenser, and bell of the instrument are enclosed in a small white enamel box, and the switch hook which projects from one side is provided with a special retaining device designed to prevent the receiver from being knocked off by the motion of the ship. The receiver is allowed to rock on the hook, otherwise the lever would lift and make a false connection when the ship was pitching and rolling.[32]

For passengers' convenience, there was also a system of 950 cabin bells to call stewards or stewardesses for

31 Lloyd's survey, 23 August 1907, LRF–PUN–W844–0024–R; *International Marine Engineering*, August 1908. *Engineering* magazine estimated that there was a total of 6,300 lamps on board, but the numbers in the text come from the August 1907 Lloyd's report, numbered by circuit.

32 *Colonist*, 18 June 1910, p.2.

assistance. There were forty indicator boards, each section of cabins having their own indicator in the companionway. Additionally, there was a master indicator at the night steward's station, which would come in handy when all the normal stewards were off-duty. There were also about forty 2½-amp electric radiators fitted in First Class staterooms, and another fifty of the 7-amp type.

Over forty electrically driven pieces of equipment could be found in the galleys. Among these were ice-cream freezers, potato peelers and dough mixers.

CRANES AND WINCHES

The four primary deck cranes, manufactured by Messrs Stothert & Pitt, were also electrically powered. Each one was capable of hoisting 3,360lb. Two of the cranes had an 18ft radius, and the other two had a radius of 26ft. They could lift at a speed of 100ft per minute, and were able to slew, or turn, at a rate of 400ft per minute. Each of the lifting motors produced 15 brake horsepower (bhp), while the separate slewing motors were rated for 2½bhp each. These cranes were mounted on the Boat Deck; the larger set was placed on the port and starboard sides of the First Class Boat Deck, just astern of the lifeboats, with their jibs often pointed aft, or stowed inboard over the Sun Deck at sea. The smaller set was mounted atop the forward portion of the Second Class Boat Deck.

These cranes were used to hoist baggage and mail on and off the ship, and were located astern because that was where most of the baggage and mails were stored. However, two electric winches were also provided on the Main Deck forward to service the fore holds; each of these winches was also rated for just over 1.5 tons. There were also two winches placed on the Forecastle by each forward cargo hatch.

All four of the aft cranes were in place when the ship left Scotland in early July 1907, and when she arrived in Liverpool, but the aft pair was removed during her first drydocking there, apparently having been deemed unnecessary. In time it was found that the second set was necessary after all, and they were later reinstalled in their original positions.

There were also four electric lifeboat winches placed atop the Sun Deck, one on each side just abaft of the No. 1 funnel, and another pair just abaft the No. 4 funnel. These winches would be used in raising and lowering the boats, and each could lift just over 5½ tons (11,200lb) at some 40ft per minute.

ELEVATORS/LIFTS

Eleven lifts, or elevators, were installed aboard *Lusitania*. Electric lifts had been in use within land-based structures since the 1880s. In 1905, the first electrically operated passenger elevator was installed aboard the Hamburg-Amerika liner *Amerika*, built by Harland & Wolff of Belfast, Northern Ireland. The decision to equip *Lusitania* with a pair of lifts in her First Class Entrance may not have been a surprise, but the inclusion was still something of a novelty on the Atlantic. The two passenger lifts were placed side by side in the central well of the Grand Staircase. The other nine lifts on the ship were primarily used for baggage, cargo or to bring foodstuffs from storage to the pantries and galleys. The table below gives more specifications on each of the eleven elevators.

Load capacity	Speed, ft/minute	Travel	Motors, rated bhp
Two passenger lifts, 1,120lb each	150	From Main Deck to Boat Deck, 36ft 3in	8
Two baggage lifts, 4,480lb each	100	From Orlop Deck to Shelter Deck	15
Two service lifts, 1,120lb each	100	From Lower Deck to Shelter Deck	5
Three food lifts, 224lb each	60	10ft to 11ft	1½
Two ash hoists, 224lb each	200	Approximately 60ft	3

The two passenger elevators serviced A (Boat) through E (Main) Decks. Each car was some 5ft 6in deep by 4ft wide. Most photographs that show the elevator spaces only show the protective grilles and glass screens that separated the stairs from the elevator and their machinery; the rarely seen cars themselves were made of solid mahogany. Each lift had two steel lift cables, each of which passed through the space between the dome over the staircase and the outer skylight, ran forward horizontally to a point past the skylight over the First Class Reading and Writing Room, and connected with the winding gear. Below the winding gear

While many photos of the entrance exist, very few are known to show either of the R. Waygood & Co. lift cars in place. This view shows the interior seat and mirror in the port-side lift, while the starboard lift remains out of sight on a lower deck. (Authors' Collection)

were the counterbalance weights. If for any reason the lift's hoist wires failed, there was a special brake to hold it in place instead of dropping it to the bottom of its shaft.

A lift call was supplied at each deck for each elevator car, with an electrical bell and indicator placed in the corresponding lift. A lift attendant, who worked a cylindrical switch with three positions – 'up', 'down' and 'stop' – saw the call on the indicator and then operated the car in response, moving it to the desired location. The handle on the switch was self-centring, so that if it was accidentally released, it would immediately revert to the centre 'stop' position, stopping the car. If the handle were to be held down as the lift neared either extreme of its track, an automatic cut-off switch would also stop the car. The gates that allowed access to the elevator could not be opened if the elevator was not in place on the other side. Another safety feature was to prevent the lift gates being left open while in operation through automated locks and electrical contacts.

VENTILATION AND CLIMATE CONTROL

Another vital system in maintaining passenger and crew comfort concerned ventilation. On the North Atlantic, liners had to contend with extremes in temperature and climate. In the summer, temperatures could easily rise into the 80s or higher outside, and in the winter it was not uncommon for temperatures to plunge well below freezing. Today we take shipboard climate control for granted, but early ships were content to provide hatches and skylights that merely allowed access to light and external air. By the time *Lusitania* was constructed, however, climate control had become more important and also more technically proficient.

Heating *Lusitania*'s interior spaces was far simpler than cooling them, since air-conditioning systems were still in their infancy and she had no true air-conditioning system in the modern sense. However, it was vital to provide adequate fresh air to even the deepest and most enclosed spaces of the ship. On a large ship like *Lusitania*, this system had to be extremely extensive. The liner adopted the 'thermo-tank' type system, which not only supplied fresh air to interior spaces, but also heated and humidified it in dry or cold conditions. In all, she was designed with forty-nine thermo-tanks – twenty-four servicing First Class areas of the ship, nine in Second Class, eleven to Third Class, and five for officer and crew accommodation.[33] Most of these thermo-tanks were mounted atop the Sun Deck; others were positioned in open-air Promenade areas. The forward thermo-tanks were placed between decks, with their fresh air supply drawn down from the aft end of the Bridge deckhouse, since forward portions of the ship could frequently be buried under solid water in heavy seas.

33 By way of comparison, *Mauretania* had fifty-three thermo-tanks, twenty-nine in First Class, nine in Second Class and fifteen for the Third Class, officer and crew spaces.

The thermo-tanks were capable of changing the air in every area they serviced, through either exhaust or supply, every seven to ten minutes. Exhausted air was carried out primarily along the Sun Deck, at the very top of the ship; drawing fresh air into the ship was done along the Promenade Deck to prevent foul discharging odours from toilets and galleys from being drawn back inside. Heating was performed by passing fresh air through hot steam tubes warmed by the boilers; under the old system, it had required three hours to heat a room to a desired temperature, but with the thermo-tank heating system it required only fifteen minutes. A temperature of at least 65°F could be maintained under even the coldest circumstances; warm air was admitted to the interior spaces through vents placed high in the rooms, so that it could heat evenly as it cooled. The cooling of interior spaces, on the other hand, was rather limited. Fresh air was drawn into the system, and brought into the interior spaces along the floor, before being drawn out through the higher vents.

There were also a dozen high-power exhaust fans connected to the galleys, lavatories and WCs; these could completely exchange the air in any given space every four minutes. A total of more than ninety heating or ventilating motors aided the ventilation of both passenger spaces and the superheated engineering spaces. Some of these fans were tied into the ventilating trunks that led to the canister-style ventilators along the Sun Deck, which have already been described.

This system may sound complete. However, when *Lusitania* entered service in September 1907, it immediately became clear that some areas of the ship were still uncomfortably warm during warm weather. Throughout her career, upgrades were made to her system of ventilation in an attempt to combat this problem. Yet this type of problem was not uncommon in ships of the period. *Olympic*, for example, was plagued with stuffy interior spaces in warm weather, also requiring upgrades throughout her career.

FIRE CONTROL SYSTEMS

It was vital to include methods of fire control in case a conflagration ever broke out aboard *Lusitania*. Due to a variety of causes – including the relatively primitive electrical wiring employed in ships of that era – it was not uncommon for even relatively new ships to suffer minor fires. Atlantic liners of the period were filled with combustible woods and veneers, and would never pass modern fire codes. It was thus important to make provisions to gain control over any such situation.

There were two complete systems of fire mains, with hose connections on each deck. Each connection was fitted within a small box, the interior of which was painted red. Also inside each box was a long leather hose, wound around a dispensing reel and fitted with a spray nozzle. These set-ups were also used by the crew to wash down the decks. There were portable chemical fire extinguishers placed in strategic locations throughout the ship. Fire alarms were also placed throughout the ship. Each alarm was triggered by a button housed behind glass, the interior of the space behind the glass being painted red for ease of identification. To detect any fires that could be lurking in less-frequented spaces of the ship – like the holds – there were also automatic fire alarms.

WATER SERVICES

Large quantities of water were needed for *Lusitania*'s boiler feed water, drinking, cooking and washing services. To meet this demand, two sets of distillers were provided. Each was capable of supplying 18,000 gallons of fresh water during any twenty-four-hour period for cooking and drinking, and of providing an additional 15,000 gallons a day for baths, toilets and washing purposes. Evaporators, not the distillers, provided 240 tons of water a day for the boilers when operating in the compound manner, or up to 250 tons a day when operating at high pressure.

Lusitania's two forward starboard-side lifeboats. The forward boat is swung out over the sea on its Stewarts and Lloyd tubular davits. Oddly, when *Lusitania* entered service, her system of numbering boats was reversed from the standard, with high-numbered boats forward and low-number boats astern. In 1907, these two boats would have been Nos. 13 and 15; later they were Nos. 1 and 3. (HFX Studios)

LIFEBOATS AND LIFE-SAVING APPLIANCES

Lusitania was believed to be 'unsinkable', a term applied by *The Shipbuilder* to her and *Mauretania*; several years later, they would use that term again in reference to *Olympic* and *Titanic*, but in that case they used the qualifying word 'practically'. For the Cunarders, no qualifying word was used. The typical thinking was that *Lusitania*, like many ships of her day, was so strong, so well subdivided into watertight compartments, that there was no chance she could ever sink.

Despite this, British Board of Trade regulations specified that *Lusitania* be endowed with lifeboat accommodations. These were the same regulations from 1894 that were still in effect when *Titanic* sank in 1912, and although *Lusitania* was not as large as the White Star liner, and did not have as high a passenger and crew capacity, her capacity still surpassed

This still from a newsreel captured on 1 May 1915 shows the lifeboat station on the starboard side of the Second Class Boat Deck. A collapsible lies beneath Boat No. 21, while further collapsibles are placed behind the station. (Authors' Collection)

those laws by a large measure. The top-end estimate considered under these regulations was for a ship of 10,000 tons; *Lusitania* was just over three times that size.

Since the regulations would not change until after the *Titanic* disaster in 1912, Cunard opted only to fit the 'unsinkable' liner with enough lifeboats to meet the regulations, not with enough to care for all passengers and crew in the event of an emergency; some sixteen lifeboats were provided when she entered service. These were mounted in four sets of four, along the fore and aft corners of the First Class Boat Deck. The boats were each 30ft long by slightly over 9ft in maximum breadth, with a depth of about 3ft 9in, although the exact width and depth dimensions of each lifeboat varied slightly.

To move these large and heavy lifeboats over the side of the ship for lowering, some thirty-two radial-type lifeboat davits were installed along the outer superstructure of the ship, one on each side fore and aft of each lifeboat. In later years, when extra davits were installed aboard *Lusitania* to handle her increased number of post-*Titanic* lifeboats, Stewarts and Lloyd's, Ltd advertised that they supplied davits to the ship; indeed, the firm supplied davits to 'many of the leading British shipbuilders'.

Lusitania's davits were described as hollow 'tubular steel' davits that were as strong as solid davits, but weighed only about half as much. These were made of mild steel, low in manganese and carbon to reduce corrosion, and were built to satisfy the survey requirements of the Board of Trade, Lloyd's and the Admiralty. The davits tapered gradually on each end rather than reducing in diameter in steps.

These radial davits were the typical style then used in most steamers. Each had their base set into a heel socket, which was riveted into the ship's side plating at the foot of B Deck. At the Boat Deck level, a collar attached to the side plating provided lateral support for the davit, and allowed it to rotate freely within its circular constraints.

The lifeboats were stowed along the Boat Deck and rested on chocks fitted to the decking. During good weather it was not uncommon for the crew to swing the boats out at least partway over the water to make more room for passengers walking the deck. To prevent damage as they swung in and bumped the corner of the deck, a boom could be fitted between the two davits. Each boat was also fitted with a snubbing chain to prevent it from swinging too far out and away from the ship.

Multiple changes were made to *Lusitania*'s lifeboat arrangement between 1912 and 1915. Collapsibles had to be fitted underneath the standard lifeboats; additional standard lifeboats were secured to the Boat Deck between the fore and aft sets, but for some time they were without matching davits. Only later were two extra sets of these tubular steel davits installed amidships on either side of the First Class Boat Deck, with another pair installed on either side of the Second Class Boat Deck astern.

When *Lusitania* started her final crossing on 1 May 1915, she had lifeboat accommodation for 2,605 – some 646 more than the official number of passengers and crew who were aboard for that final crossing – in forty-eight lifeboats. This lifeboat accommodation was broken down in the following way:

- Twenty-two wooden lifeboats – ten along either side of the First Class Boat Deck, and one on either side of the Second Class Boat Deck. Capacity: 1,323.
- Twenty-six collapsible lifeboats – eighteen stored beneath regular lifeboats, with the remaining eight stored abaft the main lifeboats. Capacity: 1,282.

She was also given thirty-five life rings, conveniently distributed throughout the ship. There were also 2,325 life jackets placed aboard, 125 of which were specifically designed for children. Many of these were placed directly in passenger cabins, with lockers on deck storing the remainder.

Below: This plan, although not specific to *Lusitania*, shows how the radial lifeboat davits of *Lusitania*'s type worked. This arrangement operated very differently from the Welin davits so familiar on *Olympic* and *Titanic*. (Bruce Beveridge Collection)

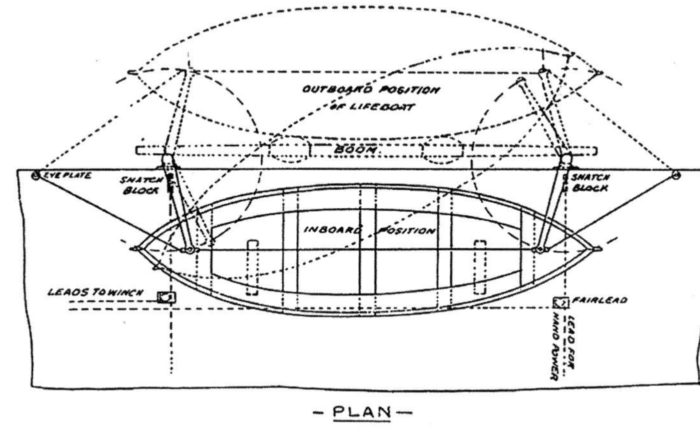

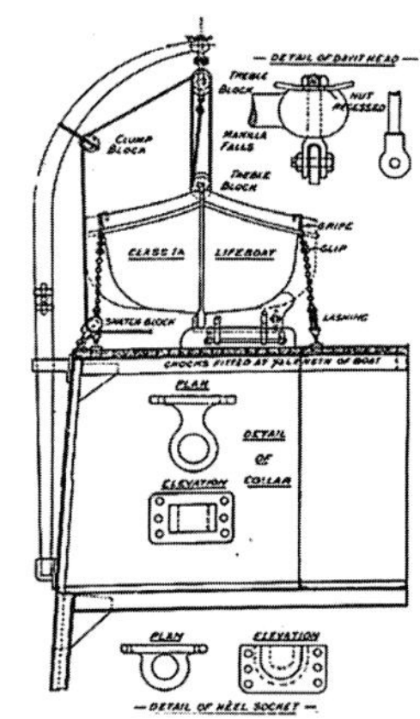

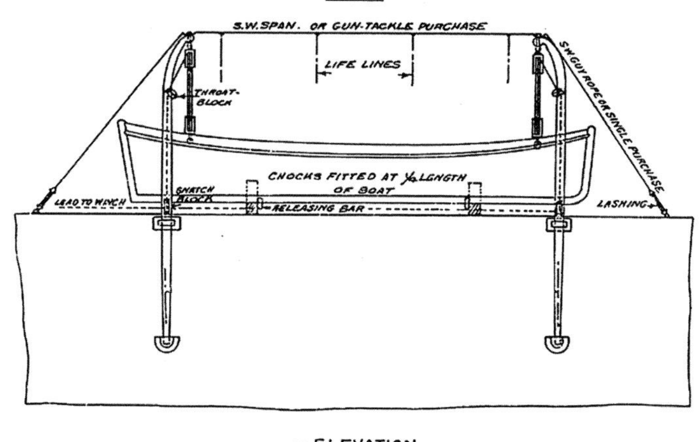

Right: Lusitania arriving in New York, *c.* 1908. The sight of the so-called 'Queen of the Seas' could make even busy New Yorkers stop everything to watch with awe and admiration. (National Archives & Records Administration, Authors' Collection)

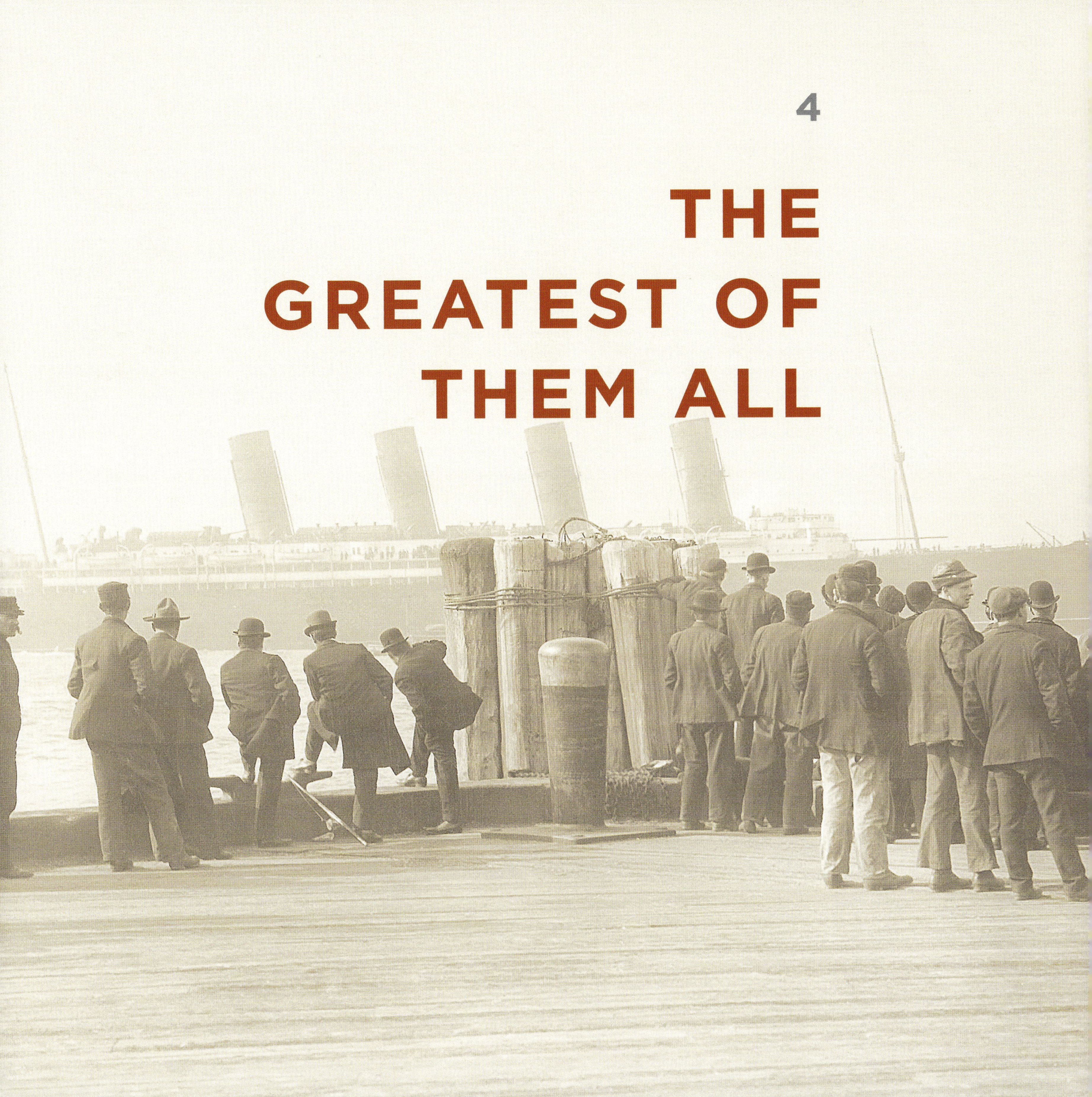

4

THE GREATEST OF THEM ALL

In early September 1907 Liverpool was the centre of a storm of excitement. *Lusitania*'s maiden voyage, arguably the most important event in the maritime community in a decade, was set to begin on Saturday, 7 September. The enormous liner sat anchored in the Sloyne, gleaming in all of her glorious newness. She was tied up in the Sloyne while the older Cunarder *Lucania* was tied up at the Prince's Stage; *Lucania* would depart Liverpool just before *Lusitania*. Since *Lucania* was the fastest British merchant ship in service before *Lusitania*, there were wild rumours that the two ships would engage in a true side-by-side race across the Atlantic to New York. Cunard tried in vain to dispel the rumours, but were, no doubt, pleased with the excitement. Not only did interest in this 'race' raise interest in the maiden voyage, but it also gave *Lusitania* a slightly lower, albeit subsidiary, goal to her primary one: recapturing the Blue Riband from Germany.

The press eagerly stirred up the public, giving Cunard and *Lusitania* more free publicity than the company could possibly have hoped for. One article lavished praise upon her in this way:

A vast ship that is so splendid a palace of luxury that one can scarcely realise that one is not in a magnificently appointed West End hotel – that is a first impression of the Cunard Company's leviathan, the *Lusitania*, the largest, most powerful, most speedy, and most wonderfully equipped vessel of her kind in the world.

Every detail of this mighty ship has been artfully designed to beguile passengers into the belief that they are on terra firma, and not crossing the wide Atlantic at all. How can one describe this wonder that has been created by the greatest geniuses of shipbuilding and decoration?

... The illusion of fashionable hotel life is carried out even to the tiniest details. One gazes out of one's cabin, not through the round port-hole of nautical tradition, but through a daintily curtained window. You turn in vain to seek the familiar bunk. Its place is taken by a comfortable, curtained brass bedstead. Arm-chairs invite you by their elegance and comfort and a shower bath is at hand, should you feel hot and

Between the end of August and 7 September 1907, *Lusitania* rested in the Sloyne in Liverpool preparing for her maiden voyage. This image is notable not only for its rarity but for its razor-sharp clarity. Canvas screens can be seen along the Promenade Deck; the ship rides at one of Cunard's new pair of buoys, but their enormous size pales beside the enormous liner's profile. (Courtesy Liverpool Record Office)

> fatigued, and the telephone is on your table ready for you to communicate instantly with any other part of the ship; or if you want to chat with fellow-passengers you may enter the lounge, the most superb piece of work on this remarkable vessel ...[1]

Cunard officials must have recognised that the public excitement and positive press bade well for a successful career. Capitalising on the public interest, on Tuesday, 3 September, *Lusitania* was opened to the public. Ferries were engaged to run people from shore to her anchorage, each paying a half-crown for the privilege of seeing the world's newest, largest and fastest ship. In about five hours, some 10,000 people visited the liner, and the proceeds were donated to charity.

Three days before the maiden voyage, Cunard representatives asked Ernest Wighton, manager of the Olympia Variety Theatre in Liverpool, if he could put together a variety show for some 800 guests on board before she entered service. Wighton accepted, took his entire company aboard, constructed an 18ft by 9ft stage in the First Class Dining Saloon, found a place for a full orchestra, and set up a 'miniature replica of our own stage performance'. Performances included a 'musical comedian', a 'musical sketch with songs and dances', and 'The three Prestons, acrobats and humourists'. The programme was such a success that Cunard briefly entertained the notion of doing regular vaudeville theatre shows aboard *Lusitania* and *Mauretania* – a concept that American playwright Charles Frohman was keenly interested in.[2] Although the concept was never put into practice, the fact that the normally reserved, all-business Cunard Company had commissioned such a performance showed their excitement at having the largest, most luxurious and hopefully the fastest ship in the world back under their house flag.

Pride in the new liner was running high throughout England, but especially in her home port. There was great confidence that *Lusitania* would reclaim the Blue Riband on this maiden trip. In some respects, this seemed well-founded, based on the results of her official trials; yet others – in particular the Germans – were adopting a 'we'll see' attitude towards the event. Bookings aboard *Lusitania* were extremely high, and quite a number of notables would be aboard; reporters were eager to discover what famous names might be included among her First Class passenger list, but Cunard denied them the particulars to 'shield' those passengers from any 'inconvenience'. However, it is likely that Cunard recognised that this shroud of secrecy would help increase the sense of excitement among the press and public – and it did.

Lucania had entered service with Cunard as their fastest ship in September 1893, and was the ship the Germans had taken the Blue Riband from in 1897; although fourteen years old and less than half *Lusitania*'s tonnage, because of the excitement over *Lusitania*'s maiden voyage and the rumoured 'race' between the two, *Lucania* had some 740 First and Second Class passengers booked. For the older ship, this was a very good showing, indeed; official word from Cunard, however, stayed the course: there would be no race.

Meanwhile, there was trouble behind the scenes. *Lusitania* was not only the largest and, it was expected, the fastest vessel in the world. She was also the most expensive. As such she, and her sister *Mauretania* to follow, were the largest single risks for insurers to back. It was said:

> Although the London underwriting market is admittedly large, the brokers of the Cunard Line have found some difficulty in completing the insurances of the *Mauretania* and *Lusitania*. Under the builders' policies the *Lusitania* was insured for £1,500,000, to cover the risk of trials, etc., although the original policies on each steamer were £1,250,000. On the Cunard fleet policies the boats are valued at £800,000 each, and additional amounts are placed on what is called 'Total loss, but including excess general average and salvage charges.' The amount placed on in this way was £400,00 on each steamer. This is now being increased to £500,000, and one result of such enormous lines being placed is that the rate gets stiffer, and, whereas liners of normal values could easily be covered at 21s per cent for 12 months against total loss only, the rates of these steamers are now over 25s per cent.[3]

It was also reported that for *Lusitania*, underwriters were having 'considerable difficulty', and that 'brokers found it to be impossible to complete their slip, rates double the ordinary being asked ... So huge indeed have recent

1 'A Trip Round Ireland', *Nelson Evening Mail*, 19 September 1907, p.1.
2 From a published interview with Ernest Wighton.
3 *Poverty Bay Herald*, 5 September 1907, p.6.

insurances been that but for the co-operation of anybody of any importance in the market they could not have been effected [*sic*].'[4]

Meanwhile, preparations to feed the first batch of passengers during the crossing were well under way. There were orders for enough foodstuffs to feed an entire city. This list included, but was not limited to, '16,000 gallons of milk, 78,000 pounds of flour, 6,000 pounds of fish, 125,000 pounds of meat, 2,000 pounds of tea, 50,000 eggs, 9,000 pounds of butter, 10,000 pounds of bacon and ham, 10,000 fowls, 45,000 oysters, 1,000 pounds of turtle, 2,000 pounds of plums'. To serve these meals, there were some '18,000 pieces of silver, 11,800 knives, 50,000 pieces of crockery', and much, much more.[5] It was elaborated:

> The service of plate ... consists of 18,000 pieces. The patterns have all been specially designed and manufactured with a view to stability and to resist the hard wear inevitable on board such an immense ship. The spoons and forks are of the old English pattern. The coffee and tea pots, sugar basins, and cream ewers are in different sizes, from the usual breakfast and tea services to tete-a-tete sets for ladies' afternoon teas. The entrée dishes, vegetable dishes, cruet frames, butter coolers, sugar baskets, sauce boats, soup tureens, soufflé dishes, etc., all have had great care bestowed upon them. The beautiful dessert dishes are works of art in themselves, being of exquisite design and lined with gold. Each article is engraved with the well-known badge of the company.[6]

The ship's crew signed on and began to gather aboard; Captain Watt and his officers were no doubt enthusiastic about starting the maiden trip. Below, in the heat and noise of the Turbine Engine Room, the Starting Platform was the control centre for the ship's engines. The assemblage of wheels, gauges, dials and other controls vaguely resembled an H.G. Wells' time machine come to life, and was the direct purview of the Chief Engineer. Scots have long been recognised as wonderful engineers, and so for this Scottish-built marvel of technology, it was only proper that *Lusitania*'s Chief Engineer was Scottish. This post was given to Alexander Duncan, formerly of the 14-year-old liner *Campania*; that vessel had been the first Cunard liner to sport a twin-screw configuration, and bore a service speed of 22 knots. Now Duncan was upgrading to a much larger, faster liner, with four screws instead of two, and a completely different type of engine technology. The Renfrew native, however, proved more than competent at working *Lusitania*'s newfangled powerplant.[7]

Arthur H. Rostron had been assigned as the ship's Chief Officer for the maiden voyage. This must have been truly exciting for him. However, the day before the trip was to begin, there came a last-minute change-up: Rostron received a promotion to his first command, the 3,200-ton *Brescia*, on the Mediterranean route. Rostron must have been quite overjoyed at receiving his first captaincy, but doubtless also had mixed emotions at leaving the world's largest liner behind for such a small ship. Rostron's career would take off quickly, however; in less than five years, he would achieve fame as the Captain of the Cunarder *Carpathia*, and not long after that, he would serve as Captain of both *Lusitania* and *Mauretania*.[8]

Lusitania's crew complement was enormous, unprecedented. There were sixty-nine in the Navigation Department, including the Captain, eight officers, eight Quartermasters, three Boatswains, three Carpenters and Joiners, two Lamp-Trimmers and Yeomen, two Masters-at-Arms, two Marconi Operators, and forty Seamen. In the Engineering Department, there were some 369 men: 33 Engineer Officers, 3 Refrigerating Engineers, 192 Firemen, 120 Trimmers, and 21 Greasers. These 438 men were necessary to run the ship between her ports in Liverpool and New York. Then came the third category, those who looked after the needs of the passengers, and of other crew members; these numbered some 389: the ship's Doctor, the Chief Purser and 2 Assistant Pursers, the Chief Steward and 2 Assistant Stewards, the Chef, 2 Barbers, 28 Cooks and Bakers, 2 matrons, 10 stewardesses, 7 Postal Clerks, 2 Typists, 50 Leading Stewards, Barkeepers, and the like, and some 280

4 *The New York Times*, 3 November 1907.

5 *Dawson Daily News*, 5 November 1907.

6 *Poverty Bay Herald*, 19 October 1907, p.1.

7 *Engineering: An Illustrated Weekly Journal*, Edited by W.H. Maw and B. Alfred Raworth, Vol. LXXXIV (July–December 1907), p.170 (originally 2 August 1907 issue).

8 *The New York Times*, 11 January 1930. Previously, Rostron was identified as being *Lusitania*'s First Officer; however, photos of Rostron taken during *Lusitania*'s trials show that he had the uniform stripes of a Chief Officer, not First. The later article must have been mistaken on his intended rank.

Lusitania angles in towards the Prince's Landing Stage on the evening of 7 September 1907. The Stage is packed with people; the port anchor has been lowered, and a tug at the bow is checking her progress in closing the Stage; meanwhile, another tug at the stern pushes that end of the ship into the pier. (Authors' Collection)

Stewards. All told, the ship's complete crew complement was some 827.[9]

With the entire crew complement signed on and the ship coaled and provisioned, the working plan for sailing day was simple: *Lucania* would depart the Prince's Landing Stage at 4.30 p.m., after which *Lusitania* would tie up there, embark her passengers, and depart at 7 p.m. Time would tell whether or not all would go as planned.

Special police arrangements were made to control the large crowds expected that Saturday evening. All through the day, sightseers and spectators arrived. 'The huge floating landing stage and the river wall parades for miles down the stream on both banks were crowded with cheering spectators, who for hours waited to view the departure of the vessel', it was said.[10] This crowd eventually reached an estimated 200,000 in number.

Meanwhile, if the Cunard offices in Liverpool had been busy in the days leading up to the departure, things really came to a head that Saturday. It was said that the 'counter of the saloon [First Class] passenger department was almost unapproachable, whilst in the steerage department the crowd filled the room'.[11]

9 At times, these figures could vary; this represents *Lusitania*'s original complement.

10 *Poverty Bay Herald*, 26 October 1907, p.11.

11 *Ibid*.

The liner continues to nudge nose-first towards the pier in this image taken only shortly after the previous one. (Stuart Williamson Collection)

In the event, *Lucania* left a half-hour late, at 5 p.m., at which point the larger liner left her anchorage in the Sloyne and, under the assistance of tugs, moved north into the Mersey.[12] To save time once she was tied up, as she moved towards the Landing Stage, her Second Class Entrance gangway door on the starboard side of E Deck was thrown open, its white-painted interior surface standing in stark contrast against her black hull. She tied up at the Prince's Stage about an hour after *Lucania*'s exodus. Soon, a gangplank had been stretched from the upper level of the platform to the First Class gangway doors on the Shelter Deck.

Passengers soon began boarding. First Class passengers came aboard via their Entrance roughly amidships on E Deck, Second Class boarded astern, and Third Class forward. Once aboard, passengers – obviously awed by the *Lusitania*'s size and beauty – set off to find their way about her labyrinthine interiors. The goals were obvious: to find their assigned cabin, the Purser's Office – 'B Deck, just off the main staircase', an obliging First Class Steward might tell them – or one of the outer decks to look down on the crowds below. As dusk began to settle over the city, one by one *Lusitania*'s lights began to flicker to life; this wash of illumination bathed the spectators in the gathering darkness below her in a sea of light. Looking up at her, the new *Lusitania* presented a 'truly magnificent spectacle' to the crowds onshore.

The process of embarking passengers aboard *Lusitania* that evening was complicated by the crew's unfamiliarity with the ship and her unprecedented size. There were also visitors who boarded the ship to see her splendour for themselves, adding to the volume of human traffic. Some congestion was noted at the gangways, but things did not go too badly.

Climbing the First Class gangway that evening were many prominent people of the day. There was Richard Croker, Jr, the son of the great ex-Tammany Hall chief and former New York boss, who had retired to an estate in Ireland, and who was travelling with his wife and daughter; Robert Balfour, a prominent Scotsman, Liberal MP, prominent London businessman, and former resident of San Francisco from 1869 to 1893; the Countess of Dunmore and her daughters Lady Victoria Murray and Lady Muriel Gore Browne – prominent

12 *Ibid.*

Although this image was not taken during the maiden voyage departure, it gives some idea of how large the liner appeared as it approached the Landing Stage that evening. (J&C McCutcheon Collection)

Christian Scientists in England; architect Louis Hay and his wife; Mrs Bertha Palmer – the apparent 'inventor' of the greatly beloved brownie – and her son, Potter Palmer, Jr, prominent Chicago citizens; Mr and Mrs Cyrus H. McCormick, also of Chicago; there was also E.H. Cunard, one of Cunard's Directors, along with other Cunard VIPs; W.J. Luke, a naval architect from John Brown, aboard with other John Brown employees to ensure the success of their firm's crowning achievement; American Senator George Sutherland (Republican, Utah, 1905–17); New Jersey State Senator E.R. Ackerman (Republican, 1906–11, US Representative from New Jersey, 1919–31); Mr and Mrs Robert Goelet; a prominent reverend from Pennsylvania, Joseph L. McCabe; George Peabody; the Consul General for Romania in London, Count Ward; 82-year-old John H. Starin, former Rapid Transit Commissioner of New York and prominent New York City businessman and Democrat; George J. Capewell, Vice-President and Superintendent of the Capewell Horse Nail Company – and many, many others. There was even a group of sixteen prominent British financial journalists, on their way to Ontario, eager to see the sights of New York City on their way there. Although no one had exercised the $4,000 option of booking either the port or starboard Regal Suites – the ship's premiere accommodation – as a complete set of rooms, they were let as separate apartments.

A special correspondent from the *Tribune* also climbed the gangplank. He recalled:

> It was like the hall of the Carlton hotel – a lift shot up and down; there were dainty decorations on the high

This splendid portrait shows *Lusitania* having just tied up to the Landing Stage to take on her first batch of paying passengers. Based on a photograph of the actual event from nearly the same angle, the painting includes the likeness of the photographer who captured the scene on the left. The gangway is being extended to the ship's side in the distance. (Painting by Elang Erlangga)

> walls; pictures flourished on the staircase; ladies sat at ease in lounge chairs; there was a post office and letter bureau. The further one went the greater grew the surprises. State rooms with brass beds, wardrobes, a telephone, and washing arrangements in a special compartment leading out of each room, music rooms, smoking rooms, tea rooms, writing rooms, a dining saloon so large and lofty that tables were set on the surrounding balcony, so that the passengers dined in two tiers – it was all so splendid and on so generous a scale that one readily gave the *Lusitania* pride of place as queen of the Western Ocean.[13]

With *Lucania* departing a half-hour late, it was a given that *Lusitania*, too, would be delayed from her scheduled departure. In the end, she left just over four hours after the smaller liner, just over two hours late, at 9.10 p.m. Captain James Watt ordered the ship's whistles be sounded, telling those below that she was about to depart. Then the lines were cast off from the Stage, and crewmen on the bow and stern reeled them back up into place; they would be used later to tie the ship up in New York at the conclusion of the crossing. With the Cunard tender *Skirmisher* at her bow, *Lusitania* 'dropped away stern first from the stage up river, and was then sheered slowly out against the incoming tide'.[14]

As the gap between her hull and the Landing Stage began to expand, patriotic fervour reached the breaking point, and the crowd of 200,000 spectators burst spontaneously into the refrains of that favourite British patriotic song, 'Rule, Britannia', 'the refrain being caught up speedily by the passengers aboard the liner. At the same time all the shipping in the stream, large and small craft, saluted with shrill and hoarse soundings of their steam whistles, and *Lusitania* proudly responded with a thrilling outburst from her own powerful siren signal', temporarily drowning out the continuing strains of 'Rule, Britannia'.[15] The correspondent for *The New York Times* reported:

> The scene as she sailed was a memorable one. Fully 100,000 [*sic*] spectators lined the landing stage and the river banks in the immediate vicinity and yelled themselves hoarse as the liner gathered headway down the river, and every steamer and riverside factory for miles along the Mersey joined in the chorus of good-bys [*sic*]. The din was deafening.

No one present would ever forget the send-off that Liverpool gave *Lusitania* that night. As the liner began to disappear 'as a foreshortening glow into the night', it was termed 'a brilliant departure.[16] Even Liverpool, accustomed to the sight of the greatest ships by day and night, had never seen anything to equal it.' As the 'hurricanes of cheers' slowly faded, there was only one question on everyone's minds: would she, after all, recapture the Blue Riband for Great Britain?

Once *Lusitania* had cleared Liverpool waters and dropped off her pilot, she picked up speed, setting a course for Queenstown, Ireland. Overall, the ship was not pushed during this leg of the crossing, since it was not until she departed Queenstown that her transatlantic crossing's timing actually began; she also encountered patches of fog down the Irish Channel, between Holyhead, Wales and Tuskar Rock Light, Ireland, near Wexford Bay. That night, the spacious accommodation aboard *Lusitania* was clearly too much for some

13 *Poverty Bay Herald*, 25 October 1907, p.4.
14 *Poverty Bay Herald*, 26 October 1907, p.11.
15 *Ibid.*
16 *Ibid.*

Chief Engineer Alexander Duncan. Even now, it seems thoroughly appropriate that a Scotsman was given oversight of *Lusitania*'s engines on the maiden voyage. (Authors' Collection)

of the passengers; quickly finding themselves lost in one of her apparently endless corridors, they had to be 'rescued' by stewards and stewardesses and directed to their cabins or to the public rooms. The ship clearly felt rock solid, and predictions ran high that there would be absolutely no seasickness during the crossing, no matter what the weather.

Some passengers were clearly disappointed that the ship was not making an all-out dash to Ireland. The officers and engineering staff knew better speeds were in store, of course; some crew members hinted to passengers that they were holding everything in reserve for the next day, with their sights locked cleanly on the *Deutschland*'s speed record. 'Wait till we get to sea', one said. One of the Cunard Line officials aboard told another: 'We are out to do things.' At about 4.30 a.m. Sunday, 8 September – near Tuskar Rock, off the south-east coast of County Wexford, Ireland – *Lusitania* caught up to and passed *Lucania*. One press report stated:

> The *Lusitania* arrived here yesterday morning at 9.25 a.m., a quarter of an hour in advance of the *Lucania*, which latter had left the landing stage at Liverpool two hours ahead of the new Cunard Atlantic flyer … When the *Lusitania* loomed up on the horizon of Roches Point, with the *Lucania* some distance astern, and was viewed from the deck of the tender proceeding seawards to intercept her, it was easy to conceive what an advance she was on all of her predecessors. … She dwarfed the splendid merchant sailing vessels, battleships, and the *Lucania* beyond all recognition.[17]

Lusitania and *Lucania* anchored just off Roche's Point.[18] *Lusitania* 'made a magnificent picture' as she lay at anchor beside the older liner. The two headlands at the mouth of the harbour were covered in people, some of whom had been waiting for three or more hours to catch a glimpse of her. It was said that aboard the tender, as it 'cleared the land and came into view of the four-funnelled monster lying outside a murmur of admiration went up from those on board'. Some 200 would-be passengers had descended upon the city with high hopes that they could take passage on the liner to New York. Unfortunately, there was no room aboard the already-booked ship; they could only watch as she transferred mail from Liverpool to the tender, took on 768 bags of mail bound for America and then left.

Lucania was given the honour of leaving Queenstown first, making it out of harbour shortly before *Lusitania* and passing Daunt's Rock – the official starting point of the transatlantic passage – at about 11.35 a.m.; *Lusitania* passed Daunt's Rock at 12.10 p.m. When *Lusitania* left Queenstown in her wake, she had 486 First Class, 483 Second Class and 1,121 Third Class passengers aboard, some 2,090 in total. When combined with some 800 crew members, there were nearly 3,000 souls aboard for the crossing.

Anticipation was high for *Lusitania* to proceed at 'Full Ahead'; however, hardly had she cleared the harbour when sporadic fog closed in. Captain James Watt was visible pacing the Bridge and Bridge Wings all afternoon, and passengers lined the ship's rails, hoping to spy the older *Lucania* as they caught up with and passed her. Everyone was disappointed that the crossing's start was at a forced saunter instead of the thoroughbred pace they had expected. Happily, by 6 p.m. the fog had cleared, and Captain Watt passed word to the Engine Room for full speed.

17 *The Morning Leader*, 9 September 1907; see also *Dundee Courier*, *Exeter and Plymouth Gazette*, *Aberdeen Press and Journal*, *Western Times*, 9 September 1907.

18 There is considerable disagreement on what time *Lusitania* and *Lucania* arrived at Queenstown; this may be due to the fact that local time at Queenstown was twenty-five minutes ahead of Greenwich Mean Time [NYT + four hours thirty-five minutes, GMT minus twenty-five minutes, a variance that remained until 1916, the year after *Lusitania* sank]. Cunard officially reported that *Lucania* entered Queenstown first, at 10.20 a.m., with *Lusitania* following at 10.25 a.m., records that were reportedly given in Irish time, but which are at a sharp variance with press reports from Queenstown, which give an arrival time for *Lusitania* much earlier. Interestingly, while Cunard reported that *Lusitania* entered the harbour first, press reports from the day of the event showed that *Lusitania* had beaten the older Cunarder to the port. → The majority of press reports from Queenstown stated that *Lusitania* arrived at 9.20 a.m. and *Lucania* anchored fifteen minutes later [*Dundee Courier*, *Exeter and Plymouth Gazette*, *Aberdeen Press and Journal*, *Western Times*, 9 September 1907]; a few [the *Daily Mirror* and *The Morning Leader*, 9 September 1907] reported that *Lusitania* anchored at 9.25 a.m., and *Lucania* anchored fifteen minutes later. A cable to *The Washington Herald* stated that *Lusitania* dropped anchor at precisely 9.26 a.m., and *Lucania* dropped anchor ten minutes later. *The New York Times* correspondent aboard *Lusitania* stated she dropped anchor at 10 a.m., but later reports from *The New York Times* placed the anchor time for *Lusitania* at 9.53 a.m. (Interestingly, if the reporter first gave the time in GMT, that would equal 9.35 a.m. Irish time; if the paper later transposed the two digits in its 9.53 a.m. report, the two reports could be mutually supportive). Finally, Cunard reported *Lucania* anchored first at 10.20 a.m., and *Lusitania* second at 10.25 a.m. Irish time. Literally none of the accounts agree with each other. A detailed analysis of the matter would be of interest to pursue, but we have chosen to include two press reports from the day and will refrain from further speculation on the matter in this volume.

A colour postcard view of Roche's Point at the mouth of Queenstown Harbour, sent in 1906. *Lusitania* and *Lucania* anchored in this vicinity on the morning of 8 September, just 'a cable's length' from each other. (Authors' Collection)

That evening, the first full-dress dinner at sea was served; this was the first because the first night after leaving port was not a formal 'dress' occasion. The custom of dropping the 'dress' standard for the first and last nights of a crossing allowed passengers time to settle in from boarding and to prepare for landing, respectively. Before dinner, finely dressed passengers were still seen lining the rails on the chilly decks, hoping to see *Lucania*. In the end, the older ship did not appear before they had to go down to the Dining Saloon.

The twin elevators in the Grand Entrance were found to be well used, especially at dinnertime. The Lift Attendants, in their gold-trimmed uniforms, busily ferried well-dressed ladies and gentlemen up and down before and after dinner. One after another, passengers flocked to the Saloon, either on the main, lower level, or the upper level with its balcony overlooking the scene below. A period reference paints the scene:

> The grand saloon, tastefully decked …, forms a sight delightful to every sense of the most cultured. Not only are the viands tempting, but the paintings, the flowers, the music, the furniture, the napery, all must please the fastidious and hyper-critical. Passengers … are served beyond the dreams of the epicure of a generation ago.[19]

Indeed, the picture was a splendid display of the Edwardian era. By this time, shipboard acquaintances had begun to grow, and old friendships were being refreshed once more. The mood in the Saloon was certainly genial that evening, as First Class passengers made their grand entrance in their resplendent finery and partook of the culinary delights presented for them. Fortunately, with the ship's steadiness, there were very few cases of *mal de mer*. Although on this and subsequent nights it was noticed that there were too few Saloon Stewards to adequately care for all the passengers' needs, and service suffered a bit because of this, overall enthusiasm over the meal was high.[20]

In a novel twist, after dinner, ladies and gentlemen were allowed to smoke in the Saloon together while they listened

19 *Outward Bound*, by John Gould (Outward Bound Co., 1905/1913), pp.15–16.

20 It was subsequently recommended that an extra steward be assigned to groups of certain numbers of tables in order to care specifically for the passengers' drink requests, leaving the other stewards available to focus specifically on food, with a consequent improvement in service.

Left: While passengers dressed for dinner, the Chef and his men slaved in the galleys to prepare their fare. Here is *Lusitania's* Chef's Office. (Authors' Collection)

Below: Majesty at sea: this view looks from the C Deck balcony over the lower level of the Dining Saloon, and gives some idea of the splendid display of the Gilded Age at dinner during *Lusitania's* maiden voyage: gentlemen in white tie and tails and ladies in their finest new gowns and jewellery. (HFX Studios)

to the orchestra play from the balcony. At 8.20 p.m. the orchestra left the Saloon, but the evening was far from over. After eating a hasty dinner, they moved up to the First Class Lounge and Music Room where, at 9 p.m., they again began to serenade passengers. Meanwhile, other passengers enjoyed an after-dinner coffee or other drink in the Verandah Café, peacefully whiling away the hours as their ship sped smoothly along across the Atlantic. Gentlemen later moved on to the Smoking Room, without the accompaniment of their lady companions, to enjoy a drink, play some cards, have a good cigar and, naturally, to place wagers on the ship's pool.

The ship's pool to guess the ship's run, which would be posted at noon the following day, was a long-standing Atlantic tradition. Since all were hoping for a record crossing, the betting was particularly keen, and the higher numbers were the most popular. Some took a more middle-of-the-road number, but two men, Messrs Archibald White and Henry Doherty, took low numbers.

LUSITANIA'S ORCHESTRA AND THE ON-BOARD DAILY SCHEDULE

In the Edwardian era, it was customary for ships to have an on-board band or orchestra to provide musical entertainment for passengers. Whereas modern cruise ships have full-time cruise directors whose sole purpose is to entertain passengers every waking moment of the day, very little in the way of entertainment was set up for the liners of *Lusitania*'s day.

Mealtimes for First Class passengers on the maiden voyage were set up:

8.30–10 a.m.: Breakfast
1 p.m.: Lunch
7 p.m.: Dinner

Passengers who wished to dine at 7.30 or 8 p.m. could arrange a party and order specially prepared dinners from the head waiter, if they made the arrangements earlier than 2 p.m. 'Supper, if required,' Cunard offered, 'must be ordered before 10 o'clock.' The bars closed at 11 p.m., and the Smoking Room at 12 midnight.

During *Lusitania*'s first year of service, an observer from either the White Star Line or Harland & Wolff took passage on the liner and took notes that would certainly help the companies in the construction of their own upcoming giant liners, *Olympic* and *Titanic*.

Looking forward and to port across the upper level of the Dining Saloon. The Broadwood piano was where the orchestra played. Sitting in the beautiful Saloon, listening to the music of the ship's orchestra, 'it was difficult to believe one's self at sea', one passenger recalled. (HFX Studios)

The observer, whose identity has not survived, found many things to his liking, and also quite a few areas where he would suggest improvements. He recorded:

> The Dining Saloon arrangements are excellent, and the small tables most comfortable, and the boon of being permitted to smoke after dinner, to ladies and gentlemen alike, is great. The music is very good and much appreciated, but the band leaves the First Class Saloon at 8.10 or 8.15 to go into the Lounge at nine. This, I think, is too early, as many passengers do not dine till half past seven or eight, and they like the music. I should say the musicians should have their food before dinner, and play certainly up to half past eight, or a quarter of an hour longer than they do at present, before going into the Lounge.[21]

21 *The 'Olympic' Class Ships*, Mark Chirnside (Second Edition, 2011), Appendix 3, p.324.

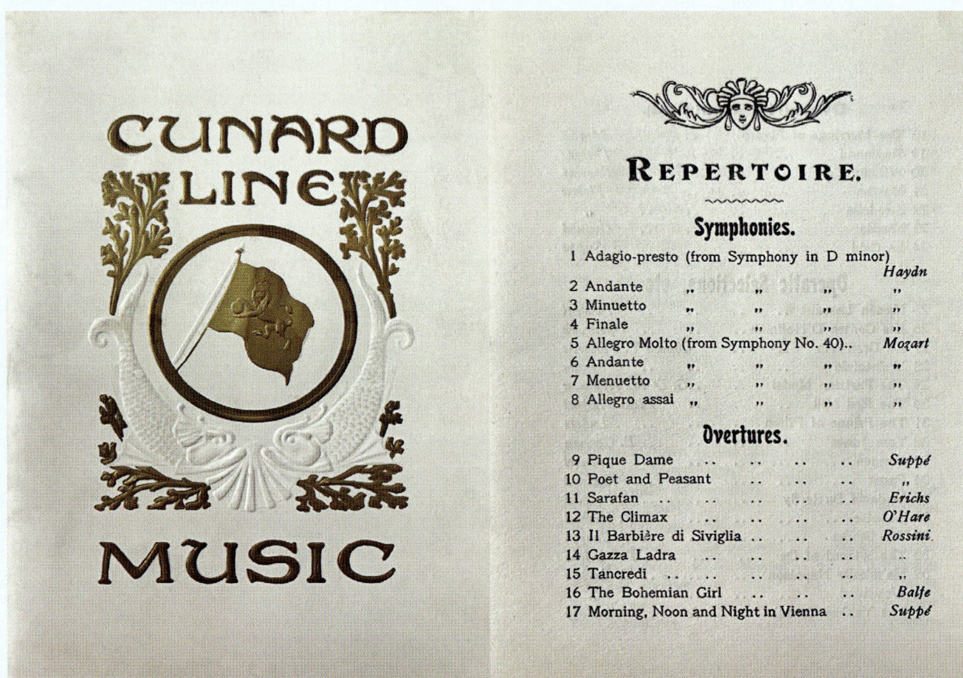

CUNARD LINE

MUSIC

REPERTOIRE.

Symphonies.

1 Adagio-presto (from Symphony in D minor) .. *Haydn*
2 Andante .. " " " "
3 Minuetto " " " "
4 Finale " " " "
5 Allegro Molto (from Symphony No. 40).. *Mozart*
6 Andante " " " "
7 Menuetto " " " "
8 Allegro assai " " " "

Overtures.

9 Pique Dame .. .. *Suppé*
10 Poet and Peasant .. .. "
11 Sarafan .. .. *Erichs*
12 The Climax .. .. *O'Hare*
13 Il Barbière di Siviglia .. .. *Rossini*
14 Gazza Ladra .. ..
15 Tancredi .. .. "
16 The Bohemian Girl .. .. .. *Balfe*
17 Morning, Noon and Night in Vienna .. *Suppé*

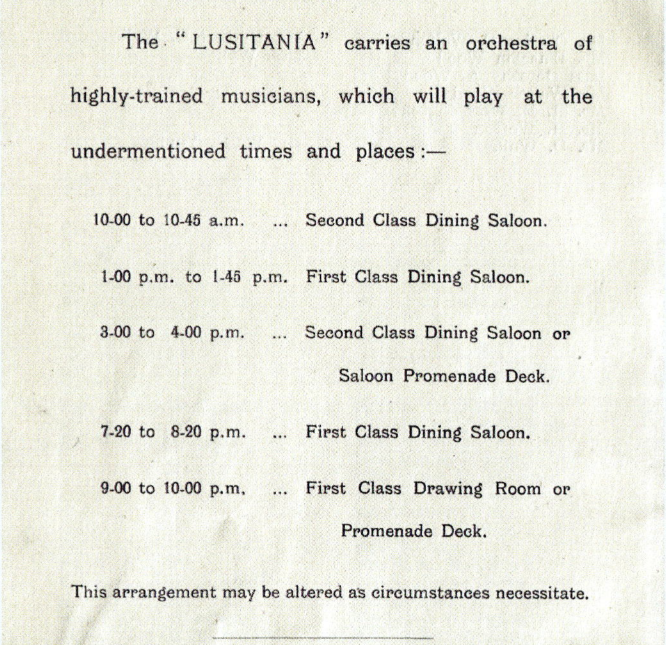

The "LUSITANIA" carries an orchestra of highly-trained musicians, which will play at the undermentioned times and places:—

10-00 to 10-45 a.m. ...	Second Class Dining Saloon.
1-00 p.m. to 1-45 p.m.	First Class Dining Saloon.
3-00 to 4-00 p.m. ...	Second Class Dining Saloon or Saloon Promenade Deck.
7-20 to 8-20 p.m. ...	First Class Dining Saloon.
9-00 to 10-00 p.m. ...	First Class Drawing Room or Promenade Deck.

This arrangement may be altered as circumstances necessitate.

Below, right: For the maiden voyage, Cunard outlined that her 'orchestra of highly trained musicians' would perform daily at the times and locations shown in this portion of the maiden voyage passenger list printed on board. (Günter Bäbler Collection)

Above: The cover and inside title page of a 1908 Cunard Line Music booklet. Hayden, Mozart, Rossini, Offenbach, Herbert, Sullivan, Puccini and Strauss were all well represented across a host of 269 symphonies, overtures, operatic selections, suites and fantasias, waltzes, intermezzos, marches, cake walks and sacred music pieces. (Authors' Collection)

The much-touted 'race' with *Lucania* was over almost before it began: the older liner's fifteen-minute lead on *Lusitania* in leaving Queenstown was erased that evening, and by midday on Monday, the new liner had a lead of 61 miles. *Lusitania*'s position at noon on Monday, 9 September was 51°7' N latitude, 23°3' W longitude. She had travelled 561 miles since 12.10 the previous day – not a match for *Deutschland*'s best day's run of 601 miles, but not a bad showing considering her cautious speed for nearly six hours due to fog. Those who had bet on high numbers the previous night were disappointed; Messrs Doherty and White made some £210 on their pessimistic bet. Hopes were still high among all, however, for a record passage.

By this time, the ship's size had simply overwhelmed some of her First Class passengers. Instead of trying to physically find their friends when they needed to get in touch with them – to ask them to dinner, to a social gathering that evening before dinner, or merely for a stroll on deck – many decided to use the novelty of their cabin telephone; the switchboard was kept busy. Reportedly, the customary British greeting, 'Are you there?' prevailed over the American, 'Hello?' for opening these conversations.

Even busier, perhaps, was the ship's pair of Marconi Operators. On average, 1,000 words a day passed through the wireless aerials suspended over the ship's quartet of tall, proud funnels. Wireless was still something of a novelty, and passengers were eager to convey information about their historic trip to friends or loved ones either ashore or on passing ships. Reporters aboard were just as eager to transmit their tales and detailed information on the ship's progress to their respective editors; they were dismayed to find that Purser Lancaster was carefully monitoring all outgoing messages from the ship to make sure no one transmitted anything negative about *Lusitania* or her speed.

On Monday, one of the reporters found this out the hard way. Lancaster scoured through the man's report to his paper, and quickly found something that he didn't like. He glanced at it severely and asked: 'What's this?' When the reporter explained, he still was not satisfied, crossed out the vexatious word and wrote it down again. The stunned reporter asked if the Purser really was the press censor. Lancaster 'drew himself dramatically up to his full height' of 5ft 1in and replied: 'I am.'

In the Smoking Room that night, after dinner, enthusiastic passengers were still optimistic for a high run; bets on 600 knots or better were taken quickly. The weather had been fine on Monday, so it was hoped that the ship's speed had been ratcheted up a bit. On the other hand, Messrs White and Doherty took the low field again, paying a mere £7.

Meanwhile, 64-year-old First Class passenger George Joseph Capewell was thoroughly enjoying his passage. Capewell had been born in Birmingham, England, but was educated in Connecticut. He was then Vice-President and Superintendent of the Capewell Horse Nail Company; seven years before, an article had reported that his 'tenacity of purpose [had] brought him to the top', and that it was Hartford,

Lusitania steams westward towards New York on a beautiful evening. Passengers' hopes are high each night that the next day's run will show improvement and fulfil expectations for a Blue Riband-winning crossing. (HFX Studios)

Every evening, the First Class Smoking Room was the scene of keen betting on the number for the daily run, which would be posted the following day just after noon. Optimists bid enthusiastically on high numbers, but on this trip it was the pessimists who tended to make a clean sweep. (Authors' Collection)

Connecticut's boast that he was one of the men who had put the city on the world map. Due to extensive business interests on both continents, Capewell was a seasoned Atlantic traveller, and had taken White Star's *Oceanic* to Europe in June; he had been able to make crossings on most of the newest Atlantic liners. Yet he was enjoying this passage more thoroughly than he recalled ever enjoying another such voyage. Capewell noticed that the waves and movement of the sea seemed to have no effect on the ship whatsoever; indeed, he was impressed to hear reports that no one had succumbed to *mal de mer*. He also found it notable that with about 3,000 people on board the liner, there was still plenty of room in the public spaces. Every detail seemed to be perfect; the ship's airy decoration was to his tastes, a welcome departure from the pompous and dark interior spaces of many of the recent German liners. He found the ship very 'light and cheerful'. Indeed, she seemed more like a 'great hotel' than a ship travelling the rugged North Atlantic at high speed. When Capewell retired at night he thought it difficult to discern, lying in his berth with no visual point of reference, whether *Lusitania* was travelling at 25 knots or standing still.

On Tuesday, one of the news correspondents on board gave Purser Lancaster – the admitted unofficial 'press censor' – a shock when he announced that he was going to send his next wireless report in code. Lancaster pored over the message with extra thoroughness, and quickly spotted something he didn't like: the phrase, 'We tank.'

'Suffering Samuel Johnson,' he exclaimed, 'what's this cryptic sentence mean?'

To this uncharacteristic behaviour on the part of a ship's Purser, the troublesome correspondent replied that it was merely his way of saying that the ship was going at 25 knots, her full speed. The explanation seems to have soothed the Purser somewhat.

This reporter's estimate of the ship's speed was misinformed, however. At noon, the day's run was calculated and posted for passengers to see: 575 nautical miles – an improvement of only 14 nautical miles over the previous day's run. Disappointment must have shown all over the ship, especially to those who had placed high-field wagers the previous evening. Messrs White and Doherty, however, must have been pleased when they pocketed £219 for their second bet. However, there was still hope for the optimists. Even at this easy pace, the gap between *Lusitania* and *Lucania* had widened to a full 120 miles, and many still felt it was possible that she could take the Blue Riband.

By this time, the routine aboard the ship was beginning to settle in. The reporters aboard noted that the *Cunard Daily Bulletin*, which was compiled by the Purser and printed each day right on board the ship, didn't have that certain 'punch' that their own newspapers ashore had. A contingent of them approached Purser Lancaster and offered to turn the *Bulletin* into a 'true' newspaper filled, they proposed, with the more interesting goings-on aboard – including the numerous shipboard romances that were already cropping up. Lancaster, somewhat surprised at the suggestion, flatly refused, and the ambitious reporters were left to find other ways to while away their time.

One made use of his time by briefly interviewing Chief Engineer Duncan. Duncan told the newspaperman that the turbines were working 'splendidly', and that 'there had not been the slightest hitch'. When asked if the ship had done her best yet, Duncan replied: 'Oh, no, we never expect to get the best results out of a ship on her maiden trip. When the engineers and stokehold staff get better acquainted with the engines and boilers we shall achieve far better results than we have done this trip.' This was a rather 'boilerplate' response from personnel of any line before a record was actually secured.

When asked whether he felt sure *Lusitania* would probably take the Blue Riband back from the Germans at some point, Duncan responded confidently: 'Perfectly sure. There's not the slightest doubt of it.' Even if the record was not recaptured on this crossing, every engineer and officer aboard *Lusitania* knew what she was capable of; taking back the Blue Riband for Britain was only a matter of time.

One reporter was able to spend quite some time in the Marconi Shack, which he described as the 'little brown cabin away on the upper deck' that few of his fellow passengers had taken note of. The cabin was very small and was absolutely packed full of gadgetry. He told his readers:

> No vessel has possessed such wonderful equipment for the transmitting of wireless messages ... [The cabin] is not big, and part of it is partitioned off for the sleeping berths of the two strenuous young electrical engineers. There is only just room to move about between brass handles, dynamos, glass tubes, mysterious wires which seem to lead nowhere, and cabinets with metal fittings containing secret and ghostly potentialities.'

Despite the cramped quarters, the two operators moved around with an 'easy swiftness'. They were kept so busy that the reporter recalled: 'During the five days I was with them, each had but four hours' sleep out of the 24. Yet, in spite of their white, tired faces, they were not merely cheerful, they were happy. "Wireless" to them had become a religion. So long as "wireless" went well, they laughed at sleep.' They listened intently as the various wireless messages came in over the headset. Sometimes the transmissions even interrupted each other. Many congratulatory messages came in, including one from *La Lorraine*.

Suddenly, at 4 p.m. on Tuesday, something else began to come through the set. The operator's clean-shaven face went 'tense with listening', and he 'flung up a warning hand' to the others. 'Quiet there, quiet,' he said. 'Here's the long-distance spark.'

The other operator, who was standing 'back among the brass handles and grim electrical coils', also gestured to the reporter for silence. Then he explained in a whisper: 'We are within two thousand miles. That is America's first word.' It was the first time that *Lusitania* had established wireless contact with the New World.

Soon enough, the operator began transmitting again, resuming 'his swift tap-tapping at the little ebony handle on his desk, filling the little chamber with an electrical crackle, like that of a coffee mill, lighting it up with blue sparks, which flashed from a glass cylinder in front of him, and told of the force which was flinging the message into the air'.

Indeed, in an era when wireless telegraphy was the cutting edge of technology, the reporter thought that the technological cocoon of the Marconi Shack was remarkable, stating that there was no other place 'so magically equipped; no place that could provide such dramatic surprises'.[22]

That same evening, there came an unpleasant change in the weather. The sea turned rough, and a strong north-north-east wind descended upon the ship. Although some rough weather had been encountered during the ship's trials, this was the first time she had met up with anything but

22 *Bush Advocate*, 23 November 1907, p.3.

smooth seas in actual service with paying customers aboard. Passengers thought the ship behaved beautifully; although the wind whistled around the outer decks, *Lusitania* was completely unaffected. Indeed, State Senator Ackerman, travelling in First Class, later said: 'On Tuesday night there was a bit of rough weather, but we only knew it because of a bulletin posted to that effect.'

At noon on Wednesday, 11 September the ship's run was again calculated and posted, this time showing some 570 knots logged – some 1,700 since leaving Queenstown; *Lusitania* had officially crossed the halfway mark of the crossing during the previous night. This run was slightly less than the previous day's, but considering the winds and seas encountered overnight, it was not a bad showing.

Not long after the run was posted, the ship's whistle sounded. Curious passengers rushed outside to see the reason for this commotion. After searching the horizon, they spotted a small steamer some distance ahead, but almost right in the giant Cunarder's path. Backing up the deafening whistle, a Scotch crewman with a powerful voice was heard to shout a warning out over the bow: 'Ship ahoy. Get out of the road, or we'll sink you.' *Lusitania* soon overtook and safely passed the smaller, significantly slower vessel, and left her in her wake.

During the remainder of Wednesday – while passengers enjoyed walking the decks, taking tea in their deck chairs and watching the sea slip by, catching up on their reading, or listening to music from the orchestra – it became clear that the liner's engines had been unleashed with a new-found vigour. Previously, little or no vibration had been felt in First Class areas; Second Class passengers may have disagreed. Yet with the ship's increase in speed, vibration was now 'perceptible in all parts of the ship'; aft spaces and the passengers occupying them would now have been suffering the full brunt of the flaws in the ship's propeller design.

That evening, the concert was held in the First Class Lounge and Music Room. This was a standard feature of every crossing on Atlantic liners of the day, and was usually held on the last night of the journey. Since the ship was scheduled to arrive Friday morning, this event would normally have taken place on Thursday night. However, with the ship's increased speed, it may have been felt that, weather permitting, the arrival might come so early Friday that a late-evening concert on Thursday might have left passengers exhausted.

At the concert, two prominent passengers took centre stage to praise *Lusitania*'s performance. During his speech, US Senator Sutherland famously coined the phrase about *Lusitania* being 'more beautiful than Solomon's Temple', as well as 'large enough to hold all his wives and mothers-in-law'. This comment raised a hearty laugh since he hailed from predominantly Mormon, and hence polygamous, Utah – although he was neither. Robert Balfour, MP also spoke about *Lusitania*, and highlighted her as yet another link of friendship between America and England.

Once the concert concluded, many passengers drifted off to their cabins for the night; some others remained awake, especially in the First Class Smoking Room, eager to make their wagers. By this point, it was the low numbers that were going quickly. Two men bid on the low number for £31, and there was some dispute as to which man had placed the wager first. The gentlemen put it to a vote and picked one of the two, while the other man used the money he had saved to buy a round of refreshments for the rest of the group. Yet other passengers were busy filling out their customs declaration forms, which were typically distributed by the stewards early on in the crossing; there were always those who procrastinated in doing this chore.

Fog moved in at 11.30 that night, forcing Captain Watt to slow his ship again. Some of the more superstitious aboard began to wonder what it was exactly that had so 'jinxed' the maiden voyage. In reality, there was nothing uncanny about it – the North Atlantic is not exactly a friendly environment for record-breaking. A good record run has to consist of two things: a powerful enough ship to make the speed and favourable weather conditions. Good weather had largely been absent. Another two hours were spent in fog on the morning of Thursday, 12 September. Because of this, when the run was posted after noon on Thursday, the ship had made only 593 nautical miles at 23.86 knots. However, it was the best run of the voyage so far. Cheers went up from enthused passengers at the marked increase in mileage. Mr Ohio Barber took £184 for a middle-of-the-road bet placed the previous evening.

With this respectable run under her belt, it still looked possible that *Lusitania* could make a record. *Deutschland*'s standing benchmark crossing, made in September 1903, was from Cherbourg, France, to New York over a course of 3,054 nautical miles. *Lusitania*'s course was 272 miles shorter so it was more important to compare the overall average speeds of the two ships. The German liner's record crossing had been made at an average speed of 23.15 knots; *Lusitania*'s speed to noon Thursday had been 23.11 knots, and her

Wednesday–Thursday run was only 8 miles short of the record 1901 run made by *Deutschland* of 601 miles. All she needed was one more good day's steaming, and the record seemed within her grasp – despite all of the foggy weather she had already encountered.

Anticipation of arriving in New York was beginning to grow; the Marconi Operators' daily regimen of 1,000 words per day suddenly doubled on Thursday. Meanwhile, in New York, eagerness was growing, as well. Reports of the ship's position began streaming in from land-based wireless stations in touch with the liner, and newspapers inked their estimates of her final time and speed. Enormous preparations were also under way to greet the ship. Extra policemen were called out in Manhattan to perform crowd control at and around the pier. Many boats and pleasure craft, large and small alike, prepared to sail out to meet her in the harbour and escort her in. Friday's early edition papers reminded the general public of the time she was due so they could be sure not to miss the event.

As it turned out, *Lusitania* covered the last 483 nautical miles without incident. She arrived at Sandy Hook, the official end-point for the crossing, at 8.05 a.m. on Friday, 13 September 1907. Her crossing had taken five days and fifty-four minutes. Her average speed for the trip had been 23.01 knots ... *Deutschland*'s record was safe by only a half-hour's time and about a tenth of a knot. Having left Liverpool with 6,500 tons of coal and arriving with just less than 1,500 tons remaining, *Lusitania* had burned roughly 5,000 tons of coal at an average rate of 41 tons every hour.

Harbour Pilot Frank Cramer and Cunard's New York General Manager Vernon Brown went out to meet the liner. Cramer came aboard via the pilot cutter *New Jersey*; as he did, someone aboard the nearby pilot cutter *New York* hoisted a sheet with the word 'Scab' on it. Ignoring this snub, Cramer climbed aboard and found his way to the Bridge, where he took up a position amidships; Cunard had specifically chosen him for this task, and he had run up and down the new Ambrose Channel several times in preparation for this momentous event. The situation was tense, and Cramer and the ship's officers were extra vigilant to make sure that everything went smoothly. Captain Watt stood directly behind Cramer, keeping an eye on the goings-on around him. Junior Third Officer Dolphin and Fourth Officer Battle were at the port and starboard Engine Room telegraphs, respectively, prepared to relay orders to Chief Duncan at a moment's notice.

A New York Harbour Pilot boarding a ship of *Lusitania*'s era. This view shows the long and perilous climb up the rope ladder that he had to undertake before arriving on deck to shepherd the ship safely to her dock. (Authors' Collection)

The ship entered the channel at a cautious 8 knots. One after another, temporary buoys came up and fell astern. As Pilot Cramer grew more confident with manoeuvring *Lusitania*'s bulk through the course, he gradually increased her speed; the ship was drawing only 30.6ft of water at the bow and 32.3ft at the stern, and the Ambrose was currently dredged to 32½ft at low tide. There was very little margin for error. Eventually, she emerged from the new cut at 12 knots, and everyone on the Bridge sighed gratefully. The next stop was Quarantine.

The revenue cutter had departed the Battery at 6.30 a.m., filled with customs men and reporters, headed for Quarantine to intercept *Lusitania*. Other boats soon showed up: *Ellis Island*, an immigration boat; *Dalzelline*, a customs cutter; *Gov. Flower*, the Health Officer's tug, with Dr Doty, the Health Officer of New York, standing outside the pilothouse in his brilliant gold-laced uniform; there was also *Manisees*, the Army Engineer's tug. Then there were the unofficial members of the welcoming flotilla: *Sirius*, a local steamboat operated by the Iron Steamboat Company, came up carrying 1,500 sightseers. As *Sirius* – some 26 years old at the time of the event – passed *Lusitania*, her passengers hurried to the starboard rails to watch the giant Cunarder go by; the small vessel leaned ominously under the imbalanced weight.

A liner's maiden reception in New York is always a proud affair; this one stood out as special, however. As *Lusitania* appeared from the light morning mists at 9.25 a.m., spectators aboard boats moving to greet her caught their breath. She was enormous, looking 'like a skyscraper adrift'. When she turned slightly to make anchorage at Quarantine, she presented her full length for the first time, and two things immediately became obvious: first, she was even bigger than everyone had at first thought; second, she was beautiful, with thoroughbred lines that put to shame the profiles of all previous Atlantic liners. Sparkling in her new paint, she looked particularly impressive as she anchored at 9.45 a.m.

Bedlam immediately erupted as every boat among the veritable flotilla in sight cut loose with their whistles, holding long, sustained blasts of salute. As *Lusitania*'s passengers lined the decks, waving back at the hospitable greeting committee, the superliner's own whistle sounded over and over again in return. The Quarantine tug approached, carrying the health inspectors who would look passengers over for signs of disease – trachoma in particular. This inspection took only half an hour. A mail boat tied up astern so that workers could begin transferring the 1,500 sacks of European mail *Lusitania* had brought. Then the revenue cutter moved up and unleashed a tide of people, including newspaper reporters, upon the liner; the reporters hurried aboard with one goal in mind: to get the story of the maiden crossing of the world's largest ship.

Once her call at Quarantine was finished, *Lusitania* began moving up the Upper Bay, towards her Lower West Side Manhattan pier. As the Statue of Liberty came up ahead, reporters moved busily from passenger to passenger, getting interviews and comments on their thoughts about the ship; these were very free with their praise of the liner, her stability and – curiously enough – her lack of vibration. All were clearly disappointed that she had not taken the Blue Riband.

Veteran North Atlantic passenger Robert Porter was just concluding his 86th crossing. He said: 'She has not done her best. ... I am certain that she is "going to do things". The fog out of Queenstown and the two hours yesterday morning caused us a delay and, of course, that could not be helped.' Robert Balfour told a reporter: 'The Captain, the chief engineer, and the naval constructor are Scotch, like myself, and so I expect great things from this steamer.' Senator Sutherland of Utah reported the liner was 'as free from vibration and as steady as it is possible for a ship to be'.

Ernest Cunard said:

> I think that the *Lusitania* is a wonderful vessel, and has more than come up to our expectations. The turbines have acted splendidly in every way. Naturally there was some slight vibration when the ship was being driven at top speed, but it was nothing like the vibration of reciprocating engines. The engines have worked without a hitch during the entire trip. No ship ever makes her record passage on her maiden voyage. The *Lusitania* averaged 25½ knots on her trial trip for forty-eight hours, and there is no reason why she should not do better later on. We are very much pleased from every point of view with the latest addition to the Cunard fleet.[23]

Indeed, he pointed out that his reason for taking passage on this crossing was 'to watch the working of the ship in order to see if we could improve the *Maretania* [*sic*], now nearing completion. I find that I shall have no suggestion to make.'

Apparently, only one thing seemed worthy of complaint at the time: New York was in the midst of a late-summer heatwave. Although the mean temperature for New York in September was then 66°F,[24] that day, the temperature moved well past 80°F. The ship's ventilation system, though extensive, was simply unable to compete. In an era when day suits and full-length dresses were de rigueur, things grew uncomfortable quickly.

23 *The New York Times*, *The New York Tribune*, 14 September 1907.
24 *The Scientific American Handbook of Travel*, p.6.

A spectacular view taken from one of the high-rises of Lower Manhattan. Looking south across the bay towards the Statue of Liberty, a flotilla of craft – described by the *New York Tribune* as a 'curious mosquito fleet' – happily escort *Lusitania* to her Manhattan pier. (National Archives & Records Administration, Authors' Collection)

This image, taken directly from the original glass-plate negative, is one of the rarest views of the well-photographed arrival of *Lusitania* in New York on her maiden voyage. Passengers throng nearly every inch of the ship's starboard-side rails; the ship leans to starboard as tugs escort her up the North River towards Pier 54. The signal flags from both masts differentiate photos of the maiden voyage arrival from all subsequent New York stays. (Syler Beaucage Collection)

While reporters were busy interviewing as many passengers as they could find, the ship continued towards Manhattan, still escorted by countless waterborne craft; the cacophony of steam whistles, shouting and cheering never ceased. As the ship neared Lower Manhattan, she looked even more impressive: she flew the US flag from her Foremast, the Cunard house flag from her Mainmast, the blue ensign of the British Royal Naval Reserve from the taffrail, and she was 'dressed' in a full display of signal flags fore and aft in celebration. The Singer Building, still under construction, was going to be the world's tallest; as *Lusitania* moved towards Manhattan, an American flag was hoisted atop it in salute. The Cunard flag on the roof of the company's offices was doffed three times, and *Lusitania*'s crew acknowledged this by lowering her own Cunard flag. From every window in Manhattan with a view of the scene, all activity slowed or ceased so that the moment could be absorbed, etched firmly in the memories of those present.

For those aboard the liner, this reception was quite a surprise. Sixty-four-year-old George Capewell thought the tops of most of the big buildings in Lower Manhattan seemed 'black with people'. As her passengers enjoyed the event, *Lusitania* moved past the Battery and up the North River. On the Bridge, Pilot Cramer brought her up past Pier 54 to a point where she could be turned around; due to the flow of the tide, this would ease her in warping into dock. For nearly an hour eight tugs pushed and pulled at her enormous weight, but eventually she was secured to the dock that would be her 'home away from home' for the next seven and a half years. Spectators on the pier and at least 5,000 more on the streets beyond waved and cheered; the scene was, initially at least, utter pandemonium. Some of the more determined individuals that pushed forward were 'roughly handled' by the police as they were tossed back into the barely contained crowd. Only once the throng had begun to thin and the mounted police intervened was a measure of order restored.

With the situation quieting down, those still present could take the time to fully appreciate the great liner's size. Her bow towered over the virtually empty pier and nearly everything else in the immediate area. New York was, even then, larger than life, and the enormous *Lusitania* instantly seemed to fit in well.

Passengers slowly began to disembark. The Countess of Dunmore and her daughters, Lady Murray and Lady Browne, left in sombre mourning attire, for the Earl of Dunmore had passed away three weeks before; their garb of woe did not harmonise, however, with the overall sense of celebration. Mrs Potter Palmer – the driving force behind the invention of the brownie – and her son, Potter Jr, were tackled by reporters eager to confirm or deny rumours that she would soon marry King Peter of Serbia. In point of fact, there were rumours surrounding Mrs Palmer and numerous European heads of state at the time. Mrs Potter denied the

rumours rather wearily, and in the end, the gossip came to naught. The group of sixteen British financial journalists raced off the ship almost as soon as the gangplank had been lowered into place. They were met by automobiles and lunched at the famous Lotos Club before touring *The New York Times* building. Their train for Canada departed that very evening.

While many were disembarking, a very few VIPs and members of the press were allowed aboard, bucking the general tide as they climbed the gangplanks. Inevitably, they found their way to the Bridge, where Fourth Officer Battle explained patiently to them the details of how all the Bridge instruments worked. Meanwhile, on the pier below, two sightseers of obvious rural origin stood looking up at the ship in awe, with their jaws slack. One looked at the other and said: 'No one will believe us when we go back and tell them about this ship. And by heck, I would not have believed it either.'

Lusitania, safely tied up at her pier, towers above everything and everyone nearby. She sparkles in the sunlight as the golden letters at her prow proudly declare her name to an admiring audience. (HFX Studios)

Lusitania is greeted by a host of well-wishers and enthusiasts at Pier 54 in this very rare original glass-plate photo. A veritable flotilla of cabs and carriages wait to meet disembarking passengers; almost to the last, all of the vehicles visible are horse drawn. The signal flags in evidence during the trip to her pier remain in place, fluttering from both masts. (Syler Beaucage Collection)

After all the hype, no one could believe that *Lusitania* had failed to take back the Blue Riband. Even now it seems out of place with the promise delivered during her trials. To explain this, it has often been repeated that it was fog that prevented *Lusitania* from making a record crossing. To an extent this may be true, because fog did force her to slow outside of Queenstown, and again Wednesday night to Thursday morning. While steaming at reduced speed, the crucial half-hour was lost. However, looking at the entire crossing, *Lusitania* clearly was not driven flat out even when the weather was good. To noon Thursday, the Blue Riband had seemed within her grasp; but afterwards, she was slowed considerably so that by 8.05 Friday morning, she had averaged only 23.01 knots for the entire crossing. Why was her speed reduced so significantly on that last run?

Even before *Lusitania* departed Liverpool, Cunard's New York agents had been trying to obtain permission for her to use the new Ambrose Channel to enter New York Harbour; the Main Ship Channel was simply too winding to accommodate her unprecedented length safely. The Ambrose Channel was far straighter and more direct, but remained incomplete; permission to use this channel was only given by special exemption during the course of the maiden crossing.

Even this special permission was limited to times from dawn to dusk. If Captain Watt had maintained his higher speed of Wednesday–Thursday, *Lusitania* would have arrived much earlier at the Ambrose Lightship, pre-dawn on Friday; this would have required waiting at the Lightship until dawn, perhaps for hours.

According to Vernon Brown, Watt had been informed prior to departing Liverpool that, 'on account of tidal conditions, it was undesirable to arrive at Sandy Hook Bar until after daylight on Friday'. Indeed, on Wednesday, 11 September, Brown had sent Watt a Marconi wireless stating that he would meet the ship 'outside the Bar about 8 o'clock on Friday morning, and for the Captain to be prepared to come in through the new Ambrose Channel at 9 o'clock'. It seems that after a good day's run between Wednesday and Thursday, Captain Watt's intention was to take his intended time of arrival off Sandy Hook and work backward from that point to dictate the speed required for the rest of the crossing; the ship was not pushed for a record at any point.[25]

Although some in Germany celebrated *Lusitania*'s 'failure' to recapture the Blue Riband, and some in America and Great Britain were quick to pounce on the relatively sluggish speed of the liner when everyone was expecting her to smash all the records, naysayers were missing one very important point. If the Germans' record was safe by only half an hour when *Lusitania* had not made anything like a true attempt at the Blue Riband, there was no doubt the new English greyhound would soon return the trophy to Great Britain.

No one knew this better than the Germans, and some were sweating bullets. They tried to throw cold water over the maiden voyage by telling the press that they would probably no longer even try to build record-breaking ships, saying that there was more profit in larger moderate-speed ships, and that record-breakers were invariably uncomfortable for their passengers.

On Sunday, 15 September, Vernon Brown issued a statement mentioning that while it took time – in some cases up to five years – to bring a ship to peak performance, he predicted that 'the patience of our German friends' would not be so sorely tested with the new Cunarder. He also

An incredible view looking down from the Crow's Nest at *Lusitania*'s prow at her New York pier. The arched derrick crutch, which was black during the ship's maiden voyage departure from Liverpool, was white when she arrived in New York for the first time, having been painted during the trip. As it was torn off during the ship's third westbound crossing, all photos of the ship in New York showing this crutch were from her first or second stay in that port in late 1907. (Authors' Collection)

25 By way of comparison, on her maiden westbound crossing, 31 May–5 June 1914, *Aquitania* averaged 23.17 knots. *Aquitania* was not designed for speed like *Lusitania* was, yet she had performed better than *Lusitania* had in September 1907.

Lusitania being coaled during one of her first two stays in New York. (National Archives & Records Administration, Authors' Collection)

touched on another serious concern: coal consumption. Many felt that even if the ship could smash through all speed records, the fuel costs would be so enormous that it would negate the speed advantage. Brown was pleased to announce that *Lusitania* had arrived in New York with her bunkers only three quarters depleted – proof positive, he said, that *Lusitania* had not been driven for speed during the crossing. 'No railroad train could have been more exact in following out its itinerary than was this ship in following out her instructions,' he said.

Although in public Cunard was beaming over *Lusitania*, a few teething problems had also come to their attention. There were issues in the lavatories, for example; the stall doors had springs to hold them in place, instead of hooks, but the springs were not equal to the ship's motion in a sea. The lavatory bowls were also too shallow, allowing the contents to slop over the rim as the liner rolled. There was also some imperfection with the supply of fresh and salt water to the baths. In general, however, the issues were minor and would be put to rights without delay.

The turnaround in New York was busy for *Lusitania*'s crew ... no one had ever performed this task on a liner so big. To help ensure the work could be done in time, the ship was closed to the general public during her stay. The re-coaling process alone was enormous; simultaneously, stewards and stewardesses were busy cleaning and prepping the accommodation in order to accept a fresh batch of passengers on the eastbound return to Europe, which was set to begin on Saturday, 21 September. This eight-day window for preparations was truncated because the ship had to be ready for public inspection by specially invited VIP guests in advance of the sailing.

On Saturday, 14 September crowds of sightseers still swarmed the streets in front of Pier 54, straining to get the best view of the ship and hoping in vain to be allowed onto the pier. By Sunday, the local fruit, lemonade and souvenir vendors caught on to the opportunity, and set up along 10th Avenue near the pier. By then, the crowds were better behaved, but even so they swarmed over Pier 56 and nearby Charles Street to get good broadside views of the ship. A few who had managed to procure boarding passes observed with envy as they climbed the gangplank. Even this visitation was short-lived; after 4.30 p.m. no one but the ship's crew was allowed aboard.

On Tuesday, there was a luncheon aboard for some 400 guests of the line. They were given the grand tour, including a specially prepared lifeboat drill. Chief Engineer Duncan even allowed his precious Engine Rooms to be inspected. On Wednesday, there was a larger inspection by some 1,500 specially invited guests, who boarded via the Second Class Gangway; they were given a tour of the First Class Dining Saloon, the Regal Suites and the public rooms on the Boat Deck. A few more VIPs came aboard on Thursday. Among these was Samuel Clemens, the author and satirist better known as Mark Twain. Twain – a former riverboat pilot and the man who had coined the term 'the Gilded Age' to describe the exciting time that was then at its peak – was quite taken with *Lusitania*. With typical good humour and spontaneity, when his tour was finished, Twain said that he would tell Noah all about her when he saw him. For those who had not been able to tour the ship during her first stay in New York, Cunard officials promised that at the end of her second westbound crossing, they would give the public more opportunities to see her splendour in person.

Meanwhile, returning home to Columbus, Ohio, Mr H.M. Hubbard, Jr, gave his thoughts on the maiden voyage: 'I knew the *Lusitania* was something out of the ordinary, but I had little idea that a vessel could be constructed that would absolutely deceive passengers. If a person was blindfolded, carried on the ship and kept below during the voyage, then blindfolded again and set down in the streets of Liverpool, he would not know he were in England. That is pretty hard to understand, but it is a fact that there was no impression of traveling among the passengers.

'When eating there was no difference from eating in the best hotels in America. There was no noise, no motion – not even a shaking of the coffee in the cups even if the cups were brimfull. I thought the ship would make even better time than she did, as did all the passengers, after we left Daunt's Rock behind. The terrible heat in the stokerooms, however, was too much for the firemen, and those who learned this early profited by the pool which was made every day. With more of a crew to feed the hungry furnaces I feel sure the boat will cut the record some hours under five days. I was most impressed with the fact that the shop was like a floating town. You could stroll several blocks along the deck. Passengers were bowing to one another just like on a city street. Instead of going to staterooms at night, it would be better to say the passengers went to their homes, or apartment houses. Ocean traveling is no longer sailing – it is merely an unconscious moving over the water.'[26]

The whirlwind stay in New York came to its conclusion on Saturday, 21 September when the ship began her second crossing of the Atlantic. She was scheduled to sail at 3 p.m. Captain Watt hosted a lunch for several prominent individuals, including Ernest Cunard, J.P. Morgan – owner of the competing International Mercantile Marine Group and White Star Line – Vernon Brown, and 71-year-old US Senator Eugene Hale (Republican, Maine), the Chairman of the Committee on Naval Affairs. Hale had previously turned down a position as Secretary of the Navy, but throughout his career had done much to help build up the US Navy through appropriations.

Bookings for the crossing were heavy: 350 in First Class, 300 in second and 980 in third. *Lusitania* was taking home many of the same First and Second Class passengers whom she had brought over with her, including many reporters, Ernest Cunard and other Cunard representatives, John Brown employees and notable men involved with the steamship industry. She was also carrying Mrs Gertrude Eversfield and her 10-year-old son Charles, who were returning to England after arriving in New York only a few weeks previously, deported due to trachoma. Two other children, who did not have the eye disease, were remaining in the United States with their father, who was soon to become a US citizen; they bade farewell to their mother and brother in the Ship's Hospital before being taken ashore.

Some 2,000 spectators showed up around Pier 54 to see the liner off. To avoid confusion, Cunard refused to allow anyone onto the ship who did not hold a ticket for the crossing; hence, many farewells between passengers and their friends and relatives took place right on the pier.

Harbour Pilot Edward Young, one of the most experienced pilots in the service, had been selected to take *Lusitania* out through the Ambrose Channel. There was some concern that the liner would ground on the bottom of the new cut while departing, because her fully loaded draught was expected to be 34ft 5in. Young had been busy during the preceding week studying the navigation for the course so that he would be fully prepared. Others aboard for the trip to open water included Vernon Brown and his assistant, J.H. Walker, Captain Herbert J. Haddock of the White Star liner *Oceanic*, and two other Cunard skippers, *Pannonia*'s Captain Irving, and *Caronia*'s Captain, William Thomas Turner. Turner would, in time, command *Lusitania*.

Just before departure, there was an incident that nearly ended in serious trouble. The liner was still held fast to her pier when one of the forward lines snapped with a loud crack. With the tide coming in, the current began to push the ship's hull forward, deeper into her berth. The stern lines were then rather slack, and as the ship crept forward, two of the gangplanks connecting ship to shore began to move forward, as well. There came a scramble to keep the platforms under their shore ends. Meanwhile, the Pier Superintendent

26 *The Van Wert Daily Bulletin*, 14 September 1907.

A view likely recorded during *Lusitania*'s first departure from New York on 21 September 1907. As the ship makes her way down the Hudson, often referred to as the 'North' River, several vessels appear to be escorting her on her way. (Authors' Collection)

noticed that there were people clustered around the tops of the gangplanks; if these bridges to the shore gave way, with people standing at the gangway, there was no telling what could ensue. Showing remarkably quick thinking, he barked at the people in the gangways, warning them to get back; although there was no time to explain the situation, everyone obeyed the command promptly. The ship only ended up moving about 10ft before the stern lines grew taut and the ship came to a stop, and the gangways ended up holding fast to the ship and the shore; clearly, however, extra precautions would need to be taken to prevent future movement.

The 3 p.m. sailing time came and went before *Lusitania* was ready to depart; she was forty minutes late when she backed out into the North River. Cheers went up as the ship left her pier for the first time, and American and British flags suddenly appeared both on shore and on her decks. There were fewer waterborne craft to salute her on the way out than had greeted her the previous week. *Lusitania*'s siren sounded only once when a ferryboat threatened to cut across her path. Once her nose was pointed downstream, she was poised to get under way in earnest in a scene closely paralleling this narrative:

… Slowly, but steadily, the craft moves, as though awakened from sleep and endowed with life. Like a leviathan she quivers … Then, with majestic motion, the stately ship glides down on the broad bosom of the river, and soon the vista of the noble bay is unfolded to the wistful vision of all on deck. As our vessel passes between Governor's Island and the titanic figure of Liberty, gazing backward one gains a last view of those colossi of the metropolis, the lofty office buildings. Distance softens the outlines, and their skyline merges in a curve of grace with the grand contour of the Brooklyn Bridge. Out past the shores of Brooklyn and Staten Island; through the Narrows, past quarantine islands and Sandy Hook, away on the port side, the coast of Long Island; then a glimpse of Fire Island lighthouse, and we are on the bosom of the deep, with no more land to be seen till we reach the welcoming shores of the Old World.[27]

Here, where the New York channel waters gave way to the Atlantic, Pilot Young and the other temporary passengers disembarked. No difficulties were encountered bringing *Lusitania* out through the unfinished Ambrose cut, despite her additional draught, since the tide was very high. At 6.41 p.m., she passed Sandy Hook Lightship and took to the open sea. Ahead lay the North Atlantic Ocean and, some 2,870 nautical miles beyond, Queenstown, Ireland. Somewhere in that course, it was hoped, she would beat the eastbound record of *Kaiser Wilhelm II*. That voyage was made in June 1904 between Sandy Hook and Eddystone – some 3,112 nautical miles – over five days, eleven hours and fifty-eight minutes at an average speed of 23.58 knots. Since she had averaged just over 23 knots westbound, it was easily conceivable that she would make an average speed to recapture the Blue Riband.

Instead, Captain Watt started the crossing at a speed that, for *Lusitania*, was a downright saunter; by the time Fire Island was 50 miles in her wake, Watt had only ratcheted her speed up to 22 knots; by noon the next day, she was making only slightly better than 22½ knots. Fog was encountered on Sunday morning, but speed was maintained because it was not particularly thick. At noon, a run of 369 miles was posted. At about that time, however, the fog grew significantly denser, and the ship slowed; her speed fell below 20 knots as she passed between Cape Sable and

27 *Outward Bound*, John Gould (1905/1913), p.14.

CUNARD QUADRUPLE SCREW (TURBINE) R.M.S. "LUSITANIA"

R.M.S. "LUSITANIA."

THURSDAY, SEPTEMBER 26th, 1907.

MENU.

Puree Soubise

Salmon Trout, Dutch Sauce

Steak and Kidney Pudding

Braised Veal, Lemon Sauce

Roast Turkey, Bread Sauce

Corned Ox Tongue, Carrots

Boiled Rice

Puree of Turnips Boiled Potatoes

COLD.

Boiled Ham

Damson Tart Gelee Cunard Macaroon Torten

Saxon Pudding

Ice Cream

Cheese Dessert

Tea Coffee

This Second Class menu is from 26 September 1907, during the ship's first eastbound crossing, heading back to England on the return leg of her maiden voyage. (Authors' Collection)

Sable Island. For the remainder of Sunday and all of Monday, fog enshrouded *Lusitania*; by Monday at noon she had only covered 524 knots. Her speed remained at 19 knots through late Monday. By noon on Tuesday, 24 September, she had covered another 525 knots.

The weather had heretofore been relatively poor, but it next took a turn for the worse; on Tuesday night, the seas mounted until they were 'very heavy'. The liner rolled from port to starboard, reaching angles of 20 degrees away from vertical. Despite this heavy roll, she charged forward through the waves just as she had been designed to do – with 'the rhythmic ease of a cruising yacht'. it was observed by one passenger. Indeed, when the run was calculated at noon Wednesday, it was found that despite the poor conditions, she had made her best run of the crossing: 530 miles.

By noon on Thursday, she had only covered another 523 miles. At last, the poor weather cleared, and although temperatures were quite chilly, the sky was clear, the sea was smooth and there was only a light breeze from the east.

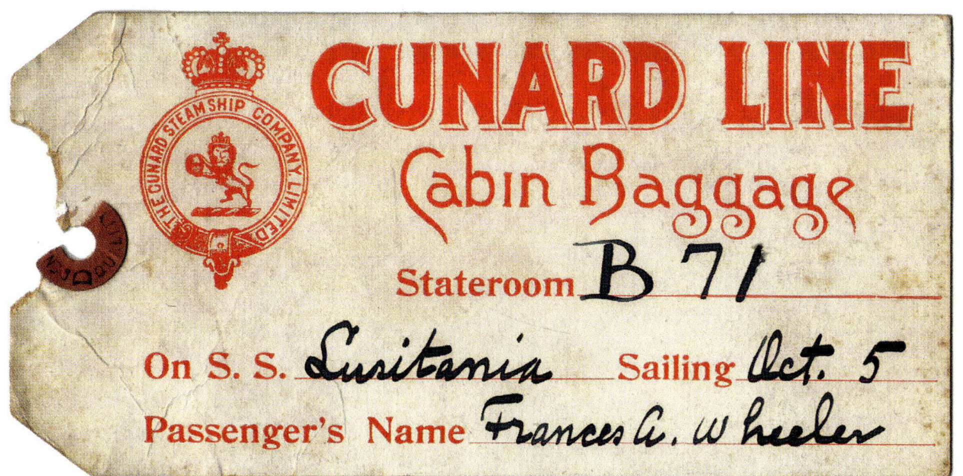

This baggage tag for passenger Frances A. Wheeler, travelling in Stateroom B71, is marked for a 5 October sailing. Comparing this to the known departure dates of *Lusitania*'s career, the only time the liner departed on that exact date was for her second westbound crossing to New York. (Stuart Williamson Collection)

But by then, it had become obvious that the transatlantic passage was nearing its end. On Thursday evening, Bull Rock off Ireland's southern coast came into view. At 1.20 on Friday morning, she passed Brow Head, another prominent Irish landmark. This signpost meant that Queenstown was only 70 miles ahead.

On Thursday evening, Captain Watt authorised a statement to be sent via wireless to the anxious press syndicates ashore. In part, it said: 'Our maiden homeward passage has been satisfactory in every respect. The machinery has behaved magnificently throughout, and has performed its full duty without a single flaw … We attempted no record on our maiden voyages except the record of landing our passengers safe and happy.' The ship covered the last 336 miles, arriving at Daunt's Rock at 3.56 a.m.; she thereupon entered the Irish port for the second time, setting up a routine that would last for much of her career. She had made the eastbound passage in five days, four hours and nineteen minutes at an average of only 22.58 knots.

Once the ship departed Queenstown Harbour on Friday morning, Captain Watt gave her a full-throttle blast of high speed. Leaving at about 7 a.m.,, *Lusitania* surged forward, covering the last 228 miles at an average speed of 25 knots and arriving in Liverpool at around 4 p.m. She landed her passengers shortly thereafter, and those proceeding to the capital by the London & North Western Rail Company arrived there just after 9 p.m., not at all an unreasonable hour.

Because neither one of *Lusitania*'s maiden crossings had taken the Blue Riband, many possible contributing factors to her 'failure' were advanced in the press: these ranged from the fog and poor weather to stoker complacency and poor-quality coal. Even so, Cunard and Captain Watt clearly never intended to drive the ship at full speed in either direction. There is simply no other way to interpret the data. However, it was also clear that in order to demonstrate that *Lusitania* was indeed the fastest ship in the world, the upcoming crossings would have to be significantly faster – they would have to recapture the Blue Riband for England, if the weather cooperated.

Behind the scenes, the Germans were desperate to find out everything they could about *Lusitania*'s engines. They had already begun to plan a response to *Lusitania* and her forthcoming sister *Mauretania*; but in order to leapfrog ahead of the Cunarders, they needed to divine everything that they possibly could about them. Many Cunard representatives had taken passage on German record-breakers when they had first made their debut on the North Atlantic, trying to learn their secrets. The Germans thus decided to send a delegation of a dozen or more engineers on *Lusitania*'s upcoming westward passage, all travelling as passengers, and all of them keen to learn about *Lusitania*'s revolutionary powerplant.

After a thorough turnaround in Liverpool between Friday, 27 September and Saturday, 5 October the great *Lusitania* was ready to embark on her next Atlantic crossing. Departing Liverpool, she made a swift 24-knot run out to Queenstown. After stopping briefly there, she took to the Atlantic with 1,994 passengers and crew aboard, only 96 fewer than had been aboard for the highly anticipated maiden voyage. She passed Daunt's Rock at 10.25 a.m. Captain Watt opened the speed queen up on the ocean for the first time; the muzzle had at last been removed, and the liner lunged forward with *Deutschland*'s record fixed firmly in her sights. Vibration astern was noted to be quite strong, a sure sign that the ship was making high speed. By noon she had logged 41 miles at a speed of 25.9 knots.

The Atlantic tossed rain, a westerly wind and heavy seas her way before Sunday evening was over, as if attempting to spoil *Lusitania*'s plans, but it did not last for long; once she had cleared the poor conditions, things remained mostly clear for the rest of the crossing, with the exception of light fog on the morning of Wednesday, 9 October. By noon on Monday, 7 October, the ship had logged another 590 miles – only 3 miles short of the best run that she had made on her maiden westbound crossing three weeks previously, and a mere 11 miles short of the *Deutschland*'s standing record.

Down below, the stokehold crew toiled in conditions that could only be called hellish. Temperatures in the four Boiler Rooms were astoundingly high; as the stokehold indicators sounded their relentless high-speed gong, the stokers opened the furnace doors and shovelled what seemed like endless quantities of coal into the furnaces before slamming the doors shut again. Wheelbarrow after wheelbarrow, shovelful after shovelful … aft, in the Engine Room, Chief Engineer Duncan watched the pressure gauges rise and kept a careful eye on the turbines as they spun at the upper ends of their designed rpms. Would the ship break the best day's record run? The crew knew that if conditions held through the night and the next morning, *Lusitania* was easily capable of it.

At noon on Tuesday, the officers fixed their position and calculated the day's run; passengers anxiously waited for it to be posted in the companionways and Smoking Rooms across the ship; they soon learned that *Lusitania* had travelled 608 miles since noon the previous day at an average speed of 24.32 knots, beating *Deutschland*'s daily record by 7 miles. Passengers were 'enthusiastic' upon receiving the news; the Blue Riband was a big step closer. There were even celebrations in the usually all-business stokeholds. A bottle of ale was given 'on the house' to each of the 323 members of the Boiler Room crew. This thoughtful generosity helped celebrate an achievement that would not have been possible without the stokers' hard work and skill.

Just as on the previous westbound crossing, betting in the ship's pool was enthusiastic. At nine o'clock each night in the Smoking Room, the auction was held and the wagers were placed. This time, however, high bidding seemed justified. On Tuesday evening, Colonel James Elverson bought the high number for $125. Overnight and throughout the next morning, *Lusitania* surged forward; the ale given to the stokers did wonders for the ship's speed, and when Tuesday's run was posted, the ship had made 617 knots at an average of 24.76 knots. Her position was 44° North latitude by 54° West longitude. Perhaps moved by the alcohol they had received the previous evening, and the hope of more to come, the stokers had managed to wring another 9 miles' worth out of the ship's machinery. Colonel Elverson took over $600 for his wager on the high number the previous night.

This second consecutive record-breaking run amply demonstrated that *Lusitania* and her crew were determined to take back the Blue Riband; it further pointed, if conditions held, to a significant lowering of that record by the time Sandy Hook was reached. Naturally, the Captain and officers all knew – based on the data from the informal and formal trials – just how fast *Lusitania* was. Now, however, it was clear to everyone that the ship was every bit as fast as promised. Looking at these figures now, one can see just how much Captain Watt had kept *Lusitania* reined in during her maiden crossings.

News of this second day of record-making was greeted in England with what could, with some reserve, be called great enthusiasm. Cunard's London offices posted an official notice advertising the remarkable 617-knot run. British newspapers speculated that *Lusitania*'s success in returning the Blue Riband to the British Empire was almost a certainty. In New York, on Thursday, Vernon Brown said that *Lusitania* still wasn't being pushed; she was, as he put it, 'just jogging along'. He denied that the liner was attempting to make a record, which was of course utter rubbish, but he did promise: 'When the Cunard officials feel the time is ripe, then you will see what the *Lusitania* can really do in the way of speed.'

If the stokers had been hoping for a second round of ale 'on the house', there must have been disappointment when none appeared. That night, rumours abounded that a drop in speed the next day would result from this regrettable lack of celebratory refreshment, and the low number went to Colonel Elverson for $175. Showing that shipboard rumours often do have a base in fact, the next run was slightly lower at 600 miles; her noon position was 41° North by 66° West, and her average speed for the entire crossing had, to that point, been at 24.12 knots.

Colonel Elverson took $836 for his wager the previous evening. One of the officers did his best to explain that the lower run had absolutely nothing to do with the supply of ale being cut off; instead, he pointed to the fact that with the coal supplies being much lower, it would take more work to get the fuel to the boilers, and that this might account for the change. Even with this slightly reduced run, it was still clear that nothing but an accident could prevent the ship from taking back the Blue Riband. Even the weather was cooperating, being clear and calm, with no further trace of fog.

At 5.25 p.m., New York time, *Lusitania* charged past the Nantucket Lightship at nearly 25 knots. Passengers lined the rails and watched with great interest as they passed the moored vessel – a sure symbol of New York's proximity – at a range of only 1¼ miles.

At dinner that evening, George Croydon Marks MP, a 49-year-old First Class passenger and representative of

On the last night of the second westbound crossing, the ship's concert was held here in the First Class Lounge. Several speeches of congratulations were made in honour of the near certainty of *Lusitania*'s impending victory by taking the Blue Riband from the German liners. (Authors' Collection)

North Cornwall, was greatly moved by *Lusitania*'s accomplishment. Since he was an engineer as well as a politician, Marks must have known just how certain it was that the liner was about to take the Blue Riband. In the great double-decked Dining Saloon, Marks gave a congratulatory speech. At one point, he turned to the dozen or so German engineers who were aboard, and pointed to them. His voice was shaking with emotion as he said: 'England once more rules the sea; and we have to thank not only Mr. Parsons, who perfected the turbine, but the Cunard Line and the British Government which financed this magnificent undertaking despite the sneers of German engineers, such as are now aboard this vessel taking notes and gathering information on the advanced science in shipbuilding.'

A little later that evening, the ship's concert was held. Donations totalling $115 were made for the firemen and stokers, a special bonus of appreciation to the crewmen who made their already record-breaking passage possible. Indeed, it was suggested by one present following the concert's conclusion that an actual stoker be called up to accept the fund; not long after, an Irish stoker named Garry appeared, mumbled a few words of thanks, and proceeded to begin lighting his pipe. Purser Lancaster quickly ordered the man below for his impudence. It was later revealed that 'Garry the Stoker' was none other than First Class passenger Bransby Williams, an actor and famous impersonator of Dickens characters. The performance, the garb, the accent – it had all been so well done that everyone, including officers who had not been told about the skit, were fooled into thinking that he was a real stoker.

There were also speeches given in honour of the powerful liner. Colonel H.I. Kowalski of San Francisco was the Chairman of the concert. He said that although the passengers applauded *Deutschland* 'as the pacemaker', they

Left: Lusitania sails up to Pier 54 for the second time in her career on the morning of 11 October 1907. She now held the Blue Riband for the fastest westbound crossing by an ocean liner. (Library of Congress, Prints & Photographs Division, Authors' Collection)

Below: This very rare abstract of a log card comes from *Lusitania*'s second westbound crossing, where she took the Blue Riband for the first time with an average speed of 24.002 knots. Interestingly, the log card does not go to the third decimal place. (Authors' Collection)

cheered *Lusitania* for 'beating her in this heroic struggle to wrest from the German giant the blue ribbon for speed. The lion is again master of the ocean highway, and this English victory is one in which the whole world joins.' He added: 'We Americans also are in on the victory.'

Colonel James Elverson Jr of Philadelphia – who in just four years' time would become the editor of *The Philadelphia Inquirer* – expressed his sentiments that *Lusitania* was a magnificent achievement. Indeed, he felt that she was the noblest and best-appointed steamship afloat. George C. Marks, MP, was also generous with his praise: 'The idea of "hands across the sea" has never been better exemplified than tonight. This trip is an international achievement, untinged with envy.' Thanks were due to the stokers, to Thomas Edison, 'the wizard of America', 'some of the most persevering of the scientists of England', and Mr Marconi, for contributing wireless communication. He also paused to 'congratulate the Cunard Company on its perseverance and indomitable pluck. It is easy to copy, but very difficult to initiate; all the world may now copy this splendid ship.'

Meanwhile, even as the concert and speeches wrapped up, the great liner continued forward on the last leg of her dash to New York. She came into sight of Fire Island at 11.22 p.m., passing the landmark at 12.07 a.m. Friday, 11 October.

ABSTRACT OF LOG OF THE

Cunard Royal Mail Steamship "LUSITANIA,"

(CAPTAIN J. B. WATT)

FROM LIVERPOOL TO NEW YORK.

Date, 1907.	Distance.	Latitude.	Longitude.	Winds.
Saturday Oct. 5	—			Left Liverpool Stage at 7-48 p.m.
Sunday ,, 6	228	To Q'stown		Arrived Queenstown at 7-45 a.m.
,, ,, ,,	41	From Daunt's	Rock L'ship.	Left Daunt's Rock 10-25 a.m.
Monday ,, 7	590	51·01 N	24·54 W	N W'ly
Tuesday ,, 8	608	48·38 ,,	40·10 ,,	S W'ly W'ly
Wednesday ,, 9	617	44·40 ,,	54·00 ,,	W'ly
Thursday ,, 10	600	41·20 ,,	66·02 ,,	W'ly N E'ly
,, ,, ,,	131	To Nantucket	Lightship.	
Friday ,, 11	193	To Sandy	Hook L'ship	1-17 a.m. Sandy Hook L'ship abeam
From Daunt's Rock L'ship to Sandy Hook L'ship	2780			AVERAGE SPEED 24·00

PASSAGE:

Daunt's Rock Lightship to Sandy Hook Lightship:--4 Days, 19 Hours, 52 Minutes.

ALL PASSENGER STEAMERS OF THE CUNARD LINE ARE FITTED WITH MARCONI'S SYSTEM OF WIRELESS TELEGRAPHY.

The final 324 miles of the voyage were completed, the ship arrived at Sandy Hook Lightship at 1.17 a.m., and *Lusitania* had made the 2,780-mile journey in four days, nineteen hours and fifty-two minutes. *Deutschland*'s previous record crossing had been at 23.15 knots; *Lusitania* had averaged 24.002 knots – over three quarters of a knot better. The

Lusitania is greeted by a host of well-wishers and enthusiasts after her victorious second arrival at Pier 54. In the lower right, a number of automobiles wait to pick up passengers, while the majority of vehicles remain horse drawn. On the ship, passengers still line the rails, looking down on the empty pier and the woefully inadequate pier structure that would soon be replaced by new facilities. (Authors' Collection)

A view of one of *Lusitania*'s first stays in New York. In the foreground, a cab or personal carriage for a wealthy passenger, driven by a pair of horses, moves forward towards the ship under the watchful eye of the driver. (Stuart Williamson Collection)

westbound Blue Riband had returned to Great Britain for the first time in nearly a decade. It was also, simultaneously, the first time that a transatlantic passage had been made in fewer than five full days; she was hailed the 'first of the four-day liners'.

Because the Ambrose Channel was still not open to use without the benefit of daylight, *Lusitania* anchored about 4 miles off Sandy Hook until, with dawn's arrival and the pilot aboard, the speedster moved up the channel; she emerged from the cut after 7 a.m. and arrived at Quarantine by 7.40 a.m. After pausing there, she resumed her journey north, reaching Pier 54 at 10.30 a.m. and tying up.

Passengers were liberal with their praise for the world's fastest ocean liner. Particularly telling, perhaps, was the fact that Mrs Isaac (Julia Barnett) Rice, who had founded

Left: One of the ship's first two stays in New York shows some interesting details: on the ship, the forward lifeboats are swung out; meanwhile, rubble and debris indicate the commencement of construction on improving the piers. What appears to be an outhouse is visible in the foreground. (Ioannis Georgiou Collection)

the 'Society for the Suppression of Unnecessary Noise' the previous year, was pleased with her voyage. Others stepping down the gangplank that morning included Senator Hale, returning from England after having gone out on *Lusitania*'s previous eastbound crossing; Viscount de Alte, Portuguese Minister to Washington; and the Honourable Claude Brabazon, a member of the British Aero Club who would soon be engaging in a twenty-four-hour balloon flight between St Louis, Missouri, and Sabina, Ohio.

Inevitably, reporters searched out Captain Watt himself. One asked: 'What do you think of the *Lusitania*?'

He responded, 'She's a daisy.'

Another question followed rapidly: 'Can she do better than this?'

With typical wit and good humour, he replied, 'Isn't this good enough?' Naturally, Watt, Chief Duncan, and others knew that *Lusitania* was capable of even more.

As promised during *Lusitania*'s maiden stay in New York, Cunard opened the ship up to visitors more freely during this turnaround. On Wednesday, 16 October a crowd of up to 15,000 persons – there were so many that an accurate

Far left: This photo was also taken during one of *Lusitania*'s first two stays in New York. Again, the view shows construction debris on shore. A clever entrepreneur has placed a placard advertising $3 trousers in the foreground. Some of the very same cobblestones visible in these early New York visit pictures are still visible today in the vicinity of Pier 54. (Ioannis Georgiou Collection)

Lusitania leaves New York on 19 October 1907 on her second eastbound crossing. As she steams past Lower Manhattan, construction work on the Singer Building progresses nicely behind her. (Authors' Collection)

headcount proved impossible – were allowed aboard. There was so much chaos around the pier that a woman reportedly fainted from overstimulation.

After staying for parts of eight days, *Lusitania* began her return trip to England at 3.30 p.m. Saturday, 19 October 1907. Three thousand gathered at her Lower Manhattan pier to see her off, cheering wildly as Captain Watt on the Bridge bowed repeatedly. She carried 429 First Class, 340 Second Class and 850 Third Class passengers, and passed Sandy Hook Lightship at 5.44 p.m. Ahead of her was the return voyage to Queenstown – and the eastbound 23.58-knot record of *Kaiser Wilhelm II*. It was now clear to everyone that *Lusitania* was capable of taking this record down. Although Captain Watt again stated he was not going to push for a record, he also allowed that if the weather held, he did not think it unlikely that she could. Below, proof positive of an upcoming attempt on the eastbound Blue Riband rested in the ship's coal bunkers: they were filled with a surplus so that she could perform at her maximum for a sustained period of time.

The weather, initially clear and fine, quickly grew poor and stayed that way for two days, with strong southeasterly gales and high head seas. Even when things eased up somewhat, the seas remained heavy and the winds remained strong. Additionally, the ship encountered heavy fog off Fastnet, Ireland. Up to noon Sunday, *Lusitania* had made 405 knots, and this was followed by runs of 570, 540 and 532 by noon Wednesday. By noon Thursday, she had added another 570 miles, leaving only 190 further miles before arriving at Daunt's Rock. Her navigation was spot on, exactly matching the 2,807-mile distance of her original eastbound crossing.

She arrived at the official end-point of the course at 9.37 p.m. Thursday, 24 October 1907. Her time over the course was four days, twenty-two hours and fifty-three minutes at an average speed of 23.61 knots – a 0.03-knot higher average than *Kaiser Wilhelm II*'s old record. *Lusitania* had beaten her own last eastbound crossing's time by five hours and twenty-six minutes, and her average speed for the entire crossing by more than a full knot. Arriving at Queenstown late Thursday evening was far preferable to arriving in the middle of the night, as she had on the last crossing. Even more astounding was the fact that all of these successes had been made despite poor weather – weather much more agitated than had been encountered on the previous eastbound trip.

LUSITANIA SONGS

Lusitania was so popular, was such a cultural phenomenon, during her early career that a number of songs were written in her honour. Among these were:

'Lusitania March And Two-Step', by F.A. Fralick (published 25 October 1907): Two-steps and rags were the music of the day, and so it was not surprising that Fralick, who lived in Toronto, chose to honour the new liner with this bouncy, cheerful piece. It was published the day after *Lusitania* had taken the Blue Riband in both directions.

'Lusitania: Queen Of The Seas Waltz', by George Manners Herd (published 5 December 1907): This piece was more genteel and classical than Fralick's two-step. It was published in London, but very little is known of its composer; it was, however, dedicated to *Lusitania*'s Captain Watt. One copy of the music, held in the Cunard Archives in Liverpool, is hand-inscribed on its cover: 'To Miss Woolner with the Composer's Compliments – George Manners Herd, Jan. 1, 1908.'

'Lusitania Waltz', by Ezra Reed (*c.* 1907): This dramatic, vibrant waltz captures attention right from its first chords, and holds listeners spellbound throughout with a deep, bassy rhythm and a brilliant melody played with the right hand that seems to float weightlessly above the bass chords. Reed had been born in Staffordshire, England, in 1862, and had been taught the piano by his older brother. He and his wife wrote over 4,000 pieces of music under 120 assumed names.

'Lusitania Intermezzo', by Alfred Rawlings (as Florence Fare) (between 1907 and 1910): Although this song bears no copyright date, it was in circulation by early 1910. Written in the key of E flat, the composer left specific instructions for playing this song, comparing it to a gavotte – a 'dainty yet courtly' French dance, with a tempo 'not too quick'. He wrote that it should be played 'lightly, but with stateliness rather than jollity'.

'Lusitania Waltzes', by Edwin E. Goffe (1910): This magnificent piece is literally a series of diverse waltzes strung together in one lengthy composition. After a brief introduction, the composer unleashes a thundering, fast-paced, recurring theme that transitions to a lilting, halting waltz theme before springing back to life again in a magnificent passage that contains a breathtaking, soaring melody. The composer was from Rhode Island, United States; very little is known of his possible motivation for publishing this piece, nearly three years after *Lusitania* entered service.

In the 2023 HFX Studios album *Liners of the Golden Age, Vol. 1*, J. Kent Layton and his wife Tessa recorded most of these fantastic, historic pieces on both modern and vintage pianos. *Liners of the Golden Age, Vol. 2* will contain a recording of '*Lusitania* Intermezzo'.

(Authors' Collection)

This ultra-rare photo was taken in Liverpool between *Lusitania*'s second and third round-trip voyages between 25 October and 2 November 1907, at a point when she held all speed records. *Mauretania* is beside her, and is about to undergo drydocking prior to her formal acceptance trials. Soon would commence a good-natured rivalry between the sisters for the Blue Riband. (Courtesy Liverpool Record Office)

Despite having already proved herself the world's largest, most luxurious and fastest ship, Captain Watt and Chief Duncan next showed that *Lusitania* had a lot more left in her. If a sense of routine was beginning to set in for Captain Watt and his officers and men, the upcoming crossing would go on to show that on the North Atlantic, nothing is ever routine.

Lusitania began her third westbound crossing on Saturday evening, 3 November 1907, with a highly respectable total of 2,037 passengers. Included were members of theatrical circles like Mrs Patrick (Beatrice) Campbell, her son Allan Urquhart and Miss Julia Marlowe. Forty-two-year-old Mrs Campbell had made her stage debut in 1888, and her first appearance on Broadway in 1902. In 1914, she would go on to play the role of Eliza Doolittle in George Bernard Shaw's play *Pygmalion*. She was a shoo-in for the role since, although she was far older than the character, Shaw had written the part of Eliza specifically for her. Julia Marlowe was aged 41 during this crossing, an English-born American actress noted for her Shakespearean portrayals, and also for her divorce in 1900 from her first husband, Robert Taber. Also aboard for the trip was Baron von Hengelmüller, Austrian Ambassador to the United States, and his wife; James B. Reynolds, Assistant Secretary of the US Treasury; and John Dunlop, the Chairman of *Lusitania*'s builders, John Brown.

Purser Lancaster also had a unique cargo to supervise: just prior to departure from the Prince's Stage, a special train had arrived from London. It carried £2,472,230 of gold – worth some $12,361,150 at the time, or about $400 million today – all 24 tons of which was going aboard *Lusitania*.[28] This was the largest single shipment of gold bullion ever carried by a liner, and the press appropriately dubbed *Lusitania* a 'treasure ship'. Most of this gold had only just arrived from South Africa the previous week for refining. Spread out over fifteen consignments, this particular shipment was going to the US Treasury and US bankers in order to back a sudden demand for currency there. This demand was so pressing that the Americans made the purchase from the British despite a currency disadvantage of almost five to one – £1 for $4.90 as of the stock market's close on Friday, 2 November to be exact.

One after another, 334 wooden boxes sealed with iron security strips were manhandled off the train and up the gangplank to the Strong Room on the Lower Deck. Purser Lancaster watched as each box came aboard and carefully ticked each, in turn, off on his list. Once the consignment was safely stowed away and under lock, Lancaster gave a receipt to the head of the bullion bankers' delegation, who had primary responsibility for the consignment's safety. With the passing of that single receipt, the Cunard Company now bore liability for the gold's safety. That accountability ultimately fell on Captain Watt, but it landed even more squarely on Purser Lancaster's shoulders. Once the ship had taken to sea, there was not an awful lot to worry about in the way of thievery; where, one asks, would a thief aboard ship go, while carrying heavy gold bars, in the middle of the Atlantic? *Lusitania* did have to reach her destination safely, however, and the gold all had to be safely unloaded on the other side. Until Lancaster could pass another receipt to those responsible for the cargo in New York, it was his charge.

The Cunard 'treasure ship' made it to Queenstown by Sunday morning, departing from there at 11.30 a.m. By noon, she had logged 21 miles at 24.24 knots; by noon on Monday, she had steamed another 606 miles at 24.28 knots; on Tuesday she added 616 miles at 24.6 knots; on Wednesday 618 at 24.8 knots; on Thursday, 610 at 24.52 knots. Things seemed to be going very well. Then *Lusitania* steamed smack into the centre of a cyclone late on Thursday, 7 November. She had seen rough weather before, but this was her first 'trial by fire' in some of the

28 Inflation rates from 1907 calculated in summer 2023. Weight from *Evening Star*, 8 November 1907.

foulest weather the North Atlantic could toss her way. 'It was as though all the demons of the deep pursued us,' Captain Watt later recalled.

The worst of it came at 4 p.m. on Thursday. Steaming through seas as high as her Bridge – or roughly 60ft – *Lusitania* had buried her nose in one particular wave, only to have it crest and collapse on her forward decks. John Dunlop, of John Brown, was no doubt keenly interested in the behaviour and sturdiness of his firm's handiwork. He said that she shook from bow to stern as this sea crashed into her Bridge; it imploded three of the heavy-duty windows overlooking the Forecastle, also twisting off the arched steel cargo boom rest on the bow; it tossed spindrift higher than the forward funnel. On the Bridge, the First Officer was cut by flying glass, but fortunately his injuries were not severe. Although the overall situation was very trying, *Lusitania* reached every expectation for behaviour. Later reports that members of the crew toiling in the Boiler and Engine Rooms would never have guessed there was a storm may have been slightly exaggerated; yet thanks to their hard work, the ship maintained an average speed of over 22 knots despite this punishment.

The Thursday concert was reportedly 'a great success' in spite of the weather. Ambassador Hengelmüller presided, and Mrs Campbell and Miss Marlowe performed a rendition of the trial scene from *The Merchant of Venice*.[29] Mrs Campbell also offered a reward of £100 to the members of the Engineering Department if they could stoke the fires enough to get the ship to Sandy Hook before midnight. Through Thursday night and Friday, the ship forged determinedly forward, carrying 2,000 passengers, over 800 crew members and some $12.3 million in gold bullion.

Only after *Lusitania* passed the Nantucket Lightship on Friday did the weather begin to ease slightly. She steamed the last 310 miles at 22.09 knots, arriving at Sandy Hook at 1.40 a.m. on Friday morning, 8 November. She had averaged 24.25 knots over a 2,781-mile course in four days, eighteen hours and forty minutes. That she had cut forty-two minutes and added roughly a quarter of a knot to her previous record average speed was pleasing; that she had done this despite a southerly gale during her last full day was astounding. Although the stokers were doubtless disappointed that they had not received Mrs Campbell's reward by reaching Sandy Hook before midnight, it was no small point of satisfaction that, despite the battering, *Lusitania* came through both safely and in good time.

In New York, Thursday evening had been a marked contrast to the scenes aboard *Lusitania*. Notably, Anton Bruckner's (1824–96) unfinished 9th Symphony in D Minor had debuted at Carnegie Hall, performed by the Boston Symphony Orchestra. It was a fitting way to open their concert season, and was well received. Yet on Friday morning,

This fantastic sketch shows the scene as $2.4 million in gold bullion is being loaded aboard *Lusitania* at Liverpool just before her sailing for the United States. (Authors' Collection)

29 Some papers reported the concert was cancelled, but details of the performance were given in *The Washington Herald*, 8 November 1907.

the excitement shifted, once again focusing on the mighty *Lusitania* as she pulled into her West Side Pier.

Burly longshoremen moved towards the liner, climbed the freight gangplank, and headed towards the Strong Room. There, Purser Lancaster had a brief conference with the American representatives of the consignors before unlocking the doors to allow for offloading. Two men took each box, straining under the great weight as they went down the gangplank. The Purser, now standing on the pier, dutifully checked off each box as they went; two others backed him up to ensure an accurate count. A special fence-enclosed portion of the pier, close to the river, was set aside as a temporary Landing Stage for the gold. Spectators on the pier lined the rail, watching as the largest single shipment of gold ever carried aboard an ocean liner was unloaded.

Once the entire shipment had been discharged, and the count tallied and double-checked, Purser Lancaster signed the second receipt, passing responsibility for the gold back to the consignors. The gold was next moved from the pier to waiting trucks, and would subsequently be taken to the Sub-Treasury and the Assay Office for distribution. The 'treasure ship' reverted to the mere mortal status of 'world's largest and fastest'. Visibly sweating although he had not been doing any of the manual labour, Purser Lancaster wiped his brow in relief and said, 'My, I am glad that's off my hands. Now I can go to luncheon.'

Lusitania remained in New York until Saturday, 16 November; as she left that port for the third time, however, she lost the title of 'world's largest ship' in service. That very evening, *Mauretania* departed Liverpool on her maiden voyage. Immediately, the younger sister seemed destined to take *Lusitania*'s records. Not only was she larger and longer, but she was carrying a $12.9 million consignment of gold, slightly larger than her sister had just carried and setting a new record. Because of the miserable weather she encountered westbound, however – on a trip that has become infamous in the annals of maritime history – she did not immediately take the Blue Riband from *Lusitania*. That would come later.

As *Mauretania* was picking her way westward through monstrous weather and enormous seas, *Lusitania* was fighting her way east through the same storm. Yet *Lusitania* still made a very respectable speed of 23.62 knots, actually slightly faster than her second eastbound trip. When she reached Queenstown at 8.25 p.m. Thursday, 21 November, the weather was still poor, and Queenstown Harbour was raucous. Two tenders were sent to intercept *Lusitania*, but the weather was so rough that after two hours of manoeuvring to come alongside her, they had to give up. Instead, one of *Lusitania*'s own lifeboats was lowered, took on the Liverpool pilot and the Queenstown despatches, and returned to the ship. Unable to land her Irish-bound passengers and mail, she departed for Liverpool.

Lusitania had set the standard, taking the Blue Riband in each direction after only a month of service. Yet, there was reason for her crew to believe that they might be hard-pressed to keep up with *Mauretania*. During the younger sister's sea trials, she had achieved speeds higher than those realised during *Lusitania*'s trials. *Mauretania*'s crew was also quite keen to beat *Lusitania*. During the 'Mary's' trials, there were some 'exciting moments' in the stokehold. One observer recorded:

> The [*Mauretania*'s] men came from the north-east coast, and there was a powerful manifestation of rivalry between the Tynesiders and their comrades on the Clyde, who had been out with the *Lusitania* some weeks previously over the self-same course. I was in the stokehold while the *Mauretania* was running round the north end of Scotland on her way to enter the measured distance and test. The ship was running easy at the time, and a gang of eager giants of the black squad had gathered before one of the boilers which at the time was not working, and were holding forth strenuously upon the speed merits of the two ships. A stoker who hailed from South Wales, and accordingly had no interest in either riversides [*sic*], started the argument by extolling the performance of the *Lusitania* over the 1,400 miles round trip. He worked his north country comrades to such a pitch that at last one of the brawniest wielders of the stokehold shovel bawled out, 'Look you 'ere, mate, we'll lick the "Lucy", even if we bust the "Mary" to do it!' To which there was such a vociferous and enthusiastic 'Aye! Aye!' that I knew the *Mauretania* was destined to be put through her paces with a vengeance ... and there was a shake of the head which told me that it would not be their fault if the liner did not do something striking. No brawn and muscle were spared on that 48-hours run in tossing the huge shovelfuls of coal from bunker to boiler furnace; the engineer never had the least doubt about the pressure

The first in a series of six rare, home-made stereo views of *Lusitania* in 1908. Five are shown in this volume, while the sixth will appear in Volume Two. In this image, *Lusitania* is tied up at the Landing Stage in Liverpool, preparing to depart for New York. (Stuart Williamson Collection)

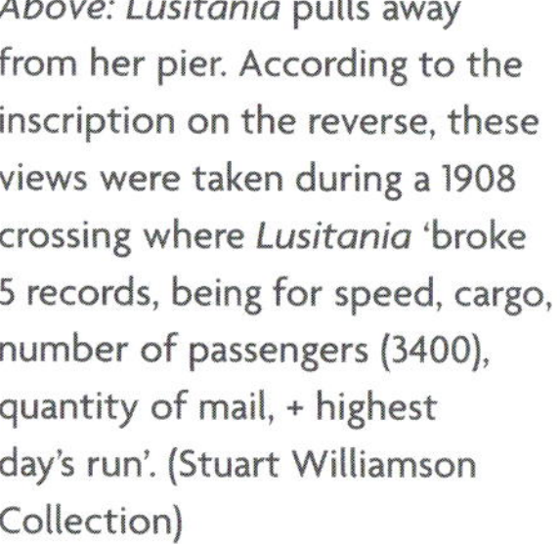

Above: Lusitania pulls away from her pier. According to the inscription on the reverse, these views were taken during a 1908 crossing where *Lusitania* 'broke 5 records, being for speed, cargo, number of passengers (3400), quantity of mail, + highest day's run'. (Stuart Williamson Collection)

Right: 'Mr Charlie Kerr' poses on the starboard Boat Deck, gripping a funnel stay for support. (Stuart Williamson Collection)

Above: A group of passengers poses on the Sun Deck just behind the No. 2 funnel. To the left are the skylight casings over the First Class Reading and Writing Room and Entrance. (Stuart Williamson Collection)

Above: Posing on the starboard Boat Deck by the aft expansion joint. The windows between the group and the photographer look in on the First Class Smoking Room. The doors behind them give entry to the starboard vestibule between the Lounge and Smoking Room. (Stuart Williamson Collection)

> of the steam within the boilers; the black squad below were seeing to that, because the pride and glory of Tyneside were at stake.[30]

Despite her failure to take the prize on her westbound maiden crossing, in returning to Europe *Mauretania* took the eastbound Blue Riband from *Lusitania*, making the passage in four days, twenty-two hours and twenty-nine minutes at 23.69 knots, just 0.07 knots faster than *Lusitania*'s fastest trip eastbound.

With that single act, *Mauretania* started a friendly rivalry between the two sisters. Now that both were in service, there grew a keen competition to see which of the two could take the record. Ultimately, it did not matter to the British – and to a lesser extent the Americans – which of the ships held the Blue Riband; more important was that the Germans were strictly out of the race. However, to many – certainly among the public, but especially to the respective crew and officers of each – which ship held the prize was very important. It was also a competition between an English-built ship and a Scottish-built one, and this tended to raise fervour between the two countries' citizens.

It has often been asked which of the two sisters was faster. The short answer is that it was *Mauretania*. The long answer is involved; perhaps a brief review of some of the records made by each ship, in turn, would help to explain this. Once *Mauretania* had taken the eastbound Blue Riband, it seemed like the North Atlantic tried everything to prevent her from taking the westbound honour; throughout the winter and early spring there were terrible storms, fog and high winds. Then, on 11 April 1908, *Mauretania* left Queenstown and, taking the long course to avoid icebergs, steamed 2,889 nautical miles to Sandy Hook in four days, twenty-three hours and fifty-nine minutes. Her average speed was 24.08 knots, and she had clipped a single minute, just sixty seconds, from *Lusitania*'s standing westbound record of five days flat made earlier that same month. *Mauretania* then held both records.

Lusitania's crew didn't take well to this development; two westbound crossings later, her nineteenth, she put up a valiant fight. First, on Wednesday, 20 May she set a daily record of 632 knots at a speed of 25.42. By the time she reached New York that Friday, her average speed was 24.83 knots, exactly three quarters of a knot faster than her upstart sister.

If *Lusitania*'s crew thought they could relax, they were in for a shock. Just a week later *Mauretania* arrived in New York at the end of her fifteenth crossing, having made a great run, recapturing the prize from her older sister. She shaved seven minutes off *Lusitania*'s time at 24.86 knots – a marginal 0.03 knots better, but astoundingly on only *three propellers*, having damaged one of her port propellers earlier that month.

Clearly something would have to be done about that, *Lusitania*'s crew must have grumbled. As bad as it was to lose the prize, it was truly humiliating to lose it to a sister suffering from such a monumental handicap. *Lusitania* retook the westbound prize on Crossing No. 23 West. When she reached Sandy Hook on 10 July she had bettered *Mauretania*'s time by thirty-nine minutes at an average 25.01 knots. It was the first time that either ship had logged an entire voyage at an average over 25 knots; *Lusitania* had simultaneously set a new best day's record run of 643 nautical miles, some 8 miles higher than *Mauretania*'s best to date. Just to make sure that *Mauretania* stayed in second place, two westbound crossings later, *Lusitania* beat her own record and brought the average speed up to 25.05 knots, logging a best day's record run of 650 nautical miles.

The situation remained rather static through autumn 1908 and the beginning of 1909; *Mauretania* was in no condition to take back *Lusitania*'s records on only three props. At the same time, *Lusitania* simply could not take *Mauretania*'s standing eastbound record. In September 1908, *Mauretania* damaged a second prop; it was promptly replaced, but the originally damaged one was not. Finally, at the end of October 1908, *Mauretania* was laid up and given entirely new outboard props; these new propellers were improved over her original screws in that they were made from a single cast. Expectation was high that the rivalry would now resume, and when *Mauretania* started her next crossing on 24 January 1909, it did.

Mauretania did not take the westbound Blue Riband right out of the gate. On her return crossing, ending 3 February, however, she lowered her own best time eastbound, bringing her average speed up to 25.2 knots. On her next westbound crossing, she took the Blue Riband back from *Lusitania*, shaving nearly two hours off her sister's best time at an average of 25.55 knots – half a knot better than

30 *Steamship Conquest of the World*, Frederick A. Talbot, (William Heinemann Co., London, 1912). Excerpt kindly provided by Art Braunschweiger.

Right: Lusitania in the Mersey early in her career, around 1908/09. The *Skirmisher* is approaching her port side. (Richard Smye Collection)

Below: Lusitania entering Liverpool at the end of an early crossing, passing the New Brighton Tower. Completed in 1900, the tower was over 500ft tall, and featured a ballroom beneath. Sadly, it was not maintained during the war years and was demolished between 1919 and 1921. (Courtesy Liverpool Record Office)

her older sister's best westbound speed. She had also set a new single-day record of 671 knots, 21 miles better than *Lusitania*'s best day westbound. By August 1909, *Mauretania* had again increased that record to 25.84 knots.

Next came the return salvo from the Scottish vessel. On Saturday, 28 August 1909, *Lusitania* left Liverpool on Crossing No. 59 West. After clearing Daunt's Rock at Queenstown, she set her teeth into the task of taking back the westbound Blue Riband from her sister. She was armed with a quartet of new propellers that had been fitted during July; two four-bladed single-cast brass screw props were on her outboard shafts, while two three-bladed manganese props were fitted inboard. Chief Engineer Duncan watched his equipment as the turbines drove the props into the water more and more quickly. For the first day, 'the weather was rough, with a head sea ...,' her Captain, William Turner, recalled later, '... but after that we had a fine smooth passage'. Turner took the ship across via the short course, inaugurating that run for the 1909 season; with this reduction in steaming distance, there was a possibility that the ship could land her passengers on Thursday night, rather than on Friday morning, and thus set a new benchmark – if the liner could manage it.

One after another, the daily runs were posted: to noon Sunday, 61 miles; to noon Monday, 650; Tuesday, 652; Wednesday, 651. On Wednesday, the weather deteriorated again, blowing first from the south-east before veering around all points to north-west. Captain Turner knew that this was beginning to slow the ship down – the question was, how much? Would she still be able to make a record crossing? That night it was obvious that she was steaming hard. The passengers were enthralled; at the concert that night, they unanimously adopted a resolution congratulating the Captain, Chief Engineer and crew on the record they felt sure was about to be set, noting their privilege to 'cross in the steamship which breaks the transatlantic record between Europe and the United States'.

Their optimism proved well-founded. At noon Thursday, 2 September, a 647-mile run was posted in spite of the weather. *Lusitania* triumphantly steamed the final 123 miles to the Ambrose Lightship, arriving there at 4.42 p.m.; she had plenty of time to navigate up to her pier and land her passengers by 8 p.m. She had recaptured the prize, making the 2,783-mile passage in four days, eleven hours and forty-two minutes at an average speed of 25.85 knots – just 0.01 knots higher than *Mauretania*'s newest record. The new propellers had given her the advantage she had needed – by a hair's breadth. As the victorious Cunarder

Lusitania sails past the Statue of Liberty, arriving in New York Harbour in the evening of 2 September 1909, having just taken the Blue Riband from *Mauretania* with a speed of 25.85 knots. It was a proud moment for the ship, but it was also the last time she would hold the speed prize. (HFX Studios)

steamed up New York's Upper Bay, past the Statue of Liberty – while the sun slowly set over the New Jersey skyline, illuminating the ship in a brilliant crimson evening's light – Captain Turner felt confident. Once he had docked, he told reporters that he thought that this record passage would 'keep [Captain] John Pritchard busy with the *Mauretania* for a time'.

Congratulations were offered from C.P. Sumner, the new General Agent for Cunard in the United States, and also from former General Agent Vernon Brown. Captain Turner replied to their sentiments with a sly smile, saying that he hoped to do even better in the future. Chief Engineer Duncan, beaming over his ship's triumph, refused 'with the usual canniness of a Scot' to give out actual coal consumption figures during the trip. It was said 'on good authority,' however, that she had averaged 1,050 tons a day, including coal for the auxiliaries, or about 4,725 tons for the crossing.

Even with new propellers and a new record, however, there was one detail of the passage that might have been unsettling to some: *Lusitania*'s best day's steaming on this crossing had only been 652 miles – 21 fewer than the best day westward made by *Mauretania*. This significant difference should have made it clear that even with *Lusitania*'s new props, she might be unable to keep up with her younger sister.

In the end, *Lusitania* only held the westbound Blue Riband for a week. On the Saturday after the older sister had arrived in New York, *Mauretania* left Liverpool and set her nose towards Queenstown; she departed there on Sunday, 5 September, passing Daunt's Rock at 10.15 a.m., some fifteen minutes later than *Lusitania* had on her previous crossing. The daily runs came in at 56, 653, 658, 643, 641 and 132, and *Mauretania* arrived at the endpoint for the crossing at 4.50 p.m., covering the same 2,783 miles as her sister, but shaving seven minutes off the trip at 25.87 knots – 0.02 knots better than *Lusitania*'s average. Although the weather had been more favourable than during *Lusitania*'s crossing, there was fog during *Mauretania*'s last day that held her up for about two hours. She was also reportedly dealing with a poorer quality of coal, which was apparently full of stones. It was said that when Chief Engineer Currie reported this to Captain Pritchard, he replied with dismissive determination: 'Never mind, Chief; we will beat the *Lusitania* even if we have only stones to get steam with,' and he was right. Some months later, in March 1910, and coming over the long course, *Mauretania* upped the average westbound speed again to 25.91 knots.

From September 1909, *Lusitania* was never able to take the Blue Riband from *Mauretania* in either direction. This was not because she did not attempt to do so. On Crossing No. 75 West, 19–25 March 1910, she tried to take the coveted trophy, this time sporting an even further improved set of propellers that were virtually identical to *Mauretania*'s. Earlier that same month, *Lusitania*'s Chief Officer Sandy G.S. McNeil talked to New York reporters about these props; he told them that he felt that with these new screws, *Lusitania* would prove faster than her sister. Up to noon on Thursday, 24 March, Chief Officer McNeil may have felt his point was being proved; *Lusitania* was making excellent time. However, on the last day, she hit heavy fog and hopes for a record were dashed. Also telling was the fact that her best runs of the crossing before the fog were only 656 on Wednesday and 655 on Thursday, numbers that still did not approach *Mauretania*'s best daily records.

Before the summer season was out, *Mauretania* further reduced her own westbound time, further complicating things for anyone hoping *Lusitania* could take back the record. On her Crossing No. 79 West, beginning on 10 September 1910, *Mauretania* made an average speed of 26.06 knots; she shaved ten minutes from her previous record while showing the same average speed.

Lusitania's late March 1910 crossing, during which she encountered poor weather, might sound like an unfair handicap. So what could *Lusitania* do in perfect weather? On Crossing No. 91 West, 17–22 September 1910, she tried again and this question was answered. There were no difficulties or poor weather, but when she arrived at Sandy Hook, she had only averaged 25.40 knots, a full hour and forty-four minutes slower than her own best time. Her highest run was only 647 miles. More proof came in after the turn of the New Year. On Crossing No. 99 West, in January 1911, she again tried and failed to make a record. Despite the fact that she was favoured by fair winds and smooth seas, the liner made the crossing in four days, thirteen hours and thirty-five minutes, one of her fastest long-route passages ever, but still not fast enough to take the record from *Mauretania*. Some fog had been encountered near the American coast, but the weather was reportedly not thick enough to retard her progress.

LUSITANIA OR *MAURETANIA*: WHICH WAS MORE POPULAR?

Right from the start, there was competition in the making ... not between two enemies, but between two sisters striving towards a common goal. True, *Mauretania* consistently triumphed over her sister's speed throughout their time in service together on the North Atlantic. Indeed, it seemed that *Mauretania* had bested her older Scottish-built sister in every sense save one: popularity. It has often been said that during their concurrent careers, *Lusitania* was more popular with the travelling public – despite the fact that she was the marginally smaller, slower and older of the two sisters. Maritime writers, apparently at a loss to explain the reason for her apparent edge in popularity, gave credit to her 'light' and 'airy' interior decor in First Class. For many decades, this was the final word on the subject: *Lusitania* was more popular with the travelling public and that was that.

BUT WAS SHE REALLY?

There are a number of factors that might be weighed into a consideration of a ship's popularity. There is emotional attachment, both during a ship's life and in memoriam after they have passed from the rugged surface of the sea. This is certainly true of *Titanic*, which sank on her maiden voyage and never completed a crossing of the Atlantic. Her much longer-lived sister *Olympic*, on the other hand, carried hundreds of thousands of passengers over a career that spanned parts of three decades. By comparison *Olympic* is vaguely remembered, if at all, largely as *Titanic*'s nearly identical sister.

A far more tangible way of ascertaining a ship's popularity, however, is to look at the numbers of passengers they carried. Again, there are different ways of viewing the matter. For example, would one count the total number of passengers carried during their careers, or during their concurrent years of service only? Certainly the latter would seem to be the only fair way of doing things. However, what if the ships made a different number of crossings during the same year? What if one was laid up for some time, as *Lusitania* was during the first half of 1913, while her sister continued plying the waves? Such circumstances would certainly distort the conclusion one way or another. However, if the vessels made a similar number of overall crossings during the period under examination, then the conclusions might not be so far off.

A stunning view of the two 'Ocean Monarchs', as the photo was captioned, tied up alongside each other on 14 October 1909 in the Canada Dock. It was a rare event to find the two sisters together in Liverpool. (J&C McCutcheon Collection)

So let us examine the matter from the period when each liner entered service through to the end of the year 1914. By this standard of measurement, which of the ships proved more popular? From September 1907 until the end of 1914, *Lusitania* carried 240,959 passengers. *Mauretania*, from November 1907 until the end of 1914, carried 247,971 passengers. By this method of calculation, *Mauretania* was 7,012 passengers ahead of *Lusitania*.

Some might still protest, however, saying that although *Mauretania* came into the picture two months after her sister, *Lusitania* was out of service for almost half of 1913. If she had been in service all through that time, would *Lusitania* have proved more popular? This leads us to our second method of calculation: to find an average of the number of passengers carried per crossing for each ship. What do these numbers reveal?

From September 1907 until the end of 1914, *Lusitania* carried an average passenger list of 1,242. From November 1907 until the end of 1914, *Mauretania* carried an average passenger list of 1,298. The average for *Mauretania*, then, was higher than *Lusitania*'s by some fifty-six persons.

After considering the matter very carefully, the truth becomes clear. The oft-told tale that *Lusitania* was more popular is nothing more than another legend that has been repeated time and again until it has been accepted as reality. The hard numbers do not lie, however: by both calculations, *Mauretania* was more popular than *Lusitania* during their concurrent careers on the North Atlantic, although by a small margin.[31]

31 '*Lusitania* and *Mauretania* – Perceptions of Popularity', Mark Chirnside (*The Titanic Commutator* 2008, Vol. 32, No. 184, pp.196–200); article also available on his website, www.markchirnside.co.uk. Full figures and statistics for each year of the ships' service are also available therein. My deepest thanks to Mr Chirnside, not only for ascertaining the truth of the matter through some great detective work, but also for giving permission to publish his findings in this volume.

For four more years after that attempt – nearly half of her steaming career of 202 crossings or 101 round-trip voyages – *Lusitania* never took back the prize. After Cunard reduced her speed by closing down her No. 4 Boiler Room in November 1914, she did not get another opportunity to make an attempt at the Blue Riband. In 1932, Sandy McNeil – who served aboard *Lusitania* from her fourth voyage in 1907 through to spring 1911, and who later commanded *Mauretania* between 1928 and 1931 – remembered in his memoirs that *Mauretania* 'was not a bit faster than the *Lusitania*'. After reviewing the cold hard facts, however, it is clear that McNeil was mistaken. He mentioned that after eight voyages, *Lusitania*'s propellers were improved, and that afterwards her speed improved significantly, specifically recalling three voyages of '25.89 knots with only a decimal difference'. However, by 1910, *Mauretania*'s averages had crept above that mark, continuing to rise throughout her career, and she alone held the record until 1929.

Interestingly, *Lusitania* did show that she still had a couple of remarkable tricks up her sleeve. In late summer 1911, a strike in Liverpool held up her scheduled departure by eight days – a catastrophic delay for a punctual service. She finally left on 28 August, and began a marathon of back-to-back crossings in an attempt to put her schedule on track again. She returned to Liverpool on Saturday, 9 September, holding a new record for a round-trip voyage across the Atlantic. She had managed the feat in just eleven days, twenty-three hours and forty-five minutes – twelve hours and thirty-two minutes better than *Mauretania*'s fastest round-trip record. This record was not due to any record-speed sailing, although by necessity her steaming was swift and the weather was favourable; rather it was due to an astounding thirty-two-hour and ten-minute turnaround in New York. Still behind schedule, she departed Liverpool again on Monday, 11 September after a very short stay in port, arriving in New York on Saturday, 16 September. She had managed to cross the Atlantic three times in three weeks; in New York, her engines were finally given a rest and cooldown. That December, she again made another record-breaking round trip – some twelve days and seventeen hours over the long course – although this was again due to a short turnaround rather than record speeds.

Perhaps even more astonishing was a feat that she managed in December 1912. From the morning to the afternoon of Friday, 19 December, she averaged 27 knots; during that period, she made nearly 28 knots for a full hour, turning 207rpm on her engines. Unfortunately, the high speed was not maintained throughout the entire crossing, nor even through a full day's steaming, so this remarkable feat did not give her any tangible record. Even so Captain Dow, her Commander at the time, described it as 'her best work ... since she has been a ship', adding that he still thought it possible she could make an Atlantic record.

In 1914, she snatched one record back from *Mauretania*. Eastbound, on Crossing No. 172, she established a new speed record for a single day's steaming in that direction. On 12–13 March, she covered 618 knots at an average speed of 26.7 knots, beating *Mauretania*'s best eastbound day's

Left: Fog was a frequent adversary for *Lusitania*. In this view overlooking the bow from the starboard Bridge wing, one begins to gather some idea of how difficult it could be for *Lusitania*'s officers to make safe passage with virtually no visibility. (HFX Studios)

Below: Lusitania emerges from a heavy fog bank. (HFX Studios)

steaming by 4 miles. Although Captain Dow had been confident of a new Atlantic record when he had left New York, *Lusitania* did not retake the Blue Riband – but her crew was not about to give up. *The New York Times* reported that they had 'sworn with strange oaths in West Street, New York, and Scotland Road, Liverpool, to win the blue ribbon of the Atlantic for their countrymen and popular commander'. When the war intervened only five months later, however, *Mauretania* still held the records, and *Lusitania*'s top-speed sailing days had reached their end. Having discussed the speed rivalry between *Lusitania* and *Mauretania* in such close detail, there is now no question that *Mauretania* had the advantage between 1907 and 1915; it is also clear that *Lusitania* was hard on her heels all the way. Which ship could have proved faster after the war, when both were converted to oil, is a subject that is wide open to hypothetical debate.

Speed, however, was just one aspect of *Lusitania*'s great career. On the North Atlantic, there is one primary enemy of speed: the whims of nature itself. Most often, this capricious nature was manifested in fog, or in 'heavy weather', with a rolling sea. However, there were times when capriciousness turned to outright violence. This most often took place during those infamous winter months on the North Atlantic. Some of the weather that *Lusitania* encountered during her career was downright breathtaking.

On Saturday, 30 November 1907, *Lusitania* was preparing to depart Liverpool for Queenstown and New York; unfortunately, the Mersey was closed in with thick fog, and for the entire day it proved impossible for her to tie up at the Prince's Landing Stage to embark her passengers. By that night, waiting passengers were sent to local hotels. On Sunday, at 8 a.m., the liner was finally able to tie up at the Landing Stage and take on her passengers. She had just enough time to cast off and move into the Mersey before the fog closed in again, this time so thoroughly that she couldn't cross the Bar and was forced to drop anchor. At 4 p.m. the fog lifted slightly, and over the next hour and a

Winter crossings on the North Atlantic often brought not just vicious storms and high seas, but also heavy snow. (HFX Studios)

half, under most cautious pilotage, she managed to pick her way into open water. Even then, there was no way for her to make time in the thick weather; it was Monday morning before she dropped anchor in Queenstown, already a full day behind schedule.

If Captain Watt was expecting an improvement out on the North Atlantic, he was sorely disappointed. *Lusitania* had not travelled far from Queenstown when she encountered a westerly gale. From that point forward, the trip was an absolute nightmare, with one gale after another tossing her around like a child's toy. The Dining Saloon, normally the centre of all shipboard activity, was virtually deserted. On Monday morning, a 'mountainous sea' broke inboard and did some damage to the ship. The bulwark of the forward Promenade Deck was dented, and the face of the Boat Deck deckhouse in the area of the Captain's Suite was heavily damaged. A support for one of the cargo booms was also twisted, and three of the booms went astray, taking a mast stay with them. Captain Watt was forced to stop for several hours while repairs were made by brave crewmen weathering the elements on the exposed Forecastle. Off Cape Race, she found herself in the centre of a cyclone sporting 40ft seas. Wave after wave smashed over her bow, completely inundating her forward regions and sending spray higher than her funnels. The sight awed her officers; they felt that the waves were moving in 'as though to bury' their ship. When the ship docked safely in New York on Sunday, 8 December Captain Watt said: 'The weather was the worst I have experienced in years. We had every kind of a gale.' He felt, however, that *Lusitania* had behaved 'admirably and proved her worth'. Other ships fighting their way through the furious seas that week – like the American liner *Philadelphia*, Hamburg-Amerika's *Amerika* and White Star's *Cretic* – all suffered more severe damage and also had harrowing tales to tell.

In February 1909, another astonishing storm battered *Lusitania*. Again the trouble began before she had even left Liverpool, where she had been in drydock for hull maintenance.

Lusitania encounters a storm at night, with lightning casting eerie shadows and catching the liner in silhouette. (HFX Studios)

When the time came, on Wednesday, 3 February, to bring her out of drydock, gales moved in. The Mersey was too rough for safe navigation, and it was decided that she should remain in drydock overnight. On Thursday, when the weather deteriorated instead of letting up, Cunard decided to prepare *Lucania* in the speed queen's place. On Friday, there came a last-minute reprieve, and *Lusitania* was cautiously eased out of drydock and up to the Landing Stage. She began taking on coal and cargo, but even before this process was completed, fog closed the Mersey in completely. Her frustrated crew and passengers could only wait; the weather did not allow her to depart on Crossing No. 41 until nine o'clock on Sunday morning. She fought her way to Queenstown in incredibly rough weather, and even upon arrival outside the Irish port on Sunday evening there was no sign of improvement. Captain Turner wisely decided not to enter the harbour until daylight, and dropped the starboard anchor in just over 10 fathoms, or 60ft, of water to wait. Astoundingly, at 10 p.m. that night, the 3¾in-diameter iron anchor chain parted because of the stresses imposed upon it. The port anchor was dropped in the hope that it would hold better than its counterpart had; it did, and at daylight, the ship eased into harbour. There, she took on extra coal before leaving Monday evening.

The poor weather continued through Monday night and all of Tuesday, and Turner nosed forward at a mere 14 knots. That night, the gale eased somewhat, and speed was increased until noon Wednesday, when the weather descended again with renewed vigour and a foul head sea. At five o'clock that afternoon, a tremendous wave – tall enough to stand over the upper Bridge – crashed over her bow. It struck with such force that the port-side steel crew ladder from the Promenade Deck to the Boat Deck was nipped cleanly from its bolted mounts; a piece of the teak rail on the forward Boat Deck was carved out, the rail on the port Bridge Wing was split, and the canvas windscreen and framework was torn away. The teak shutters for the Bridge windows were smashed, and the Bridge's structure was dented, as well. One First Class passenger – Andrew D. Provand, a former member of the Glasgow Parliament (Liberal) – was nearly killed; standing on the forward Boat Deck against all better judgement, he saw the wave descending on him and turned to run. He reached the companionway, but slipped as he went, cutting his forehead as he ducked inside. Despite this punishment, *Lusitania* again 'behaved splendidly'. Only three passengers required the attention of the Ship's Surgeon for seasickness.

For centuries, sailors have reported run-ins with tremendous seas. One period reference work discussed waves of 'about 40 feet' as a 'common estimate of the height of the larger waves in a severe gale on the North Atlantic'. However, it added:

> It is difficult to say what may be the greatest height of the solitary or nearly solitary waves that are from time to time reported by mariners. The casual combination of the numerous independent undulations running on the sea presumably sometimes produced two or three succeeding ridges or two or three neighbouring domes of water of considerably greater dimensions than those of the ordinary maximum waves of a storm. Although these large cumulative waves may be frequently produced, yet they will be comparatively seldom observed, because so small a fraction of the ocean's surface is at one time under observation. There are seemingly reliable accounts of cases in which these 'topping seas' have reached the height of 60 feet.[32]

In intervening decades, scientists began to scoff at these reports of 'freak', 'cumulative', 'topping' or 'rogue' waves; according to their mathematical computations, such waves could occur only once in ten millennia. In recent years, however, after incontrovertible evidence has come to light, they have been forced to re-examine their conclusions. Indeed, the 60ft estimate referred to in this period publication is not by any means the maximum height of a rogue wave, and they can occur in weather poor or fair. Sailors who survived encounters with these waves reported that they would appear out of nowhere, as if by magic.

The wave that smashed into *Lusitania* in February 1909 was most likely one of these rogue waves, but the North Atlantic had even more in store for *Lusitania*. Looking back on the events of Monday evening, 10 January 1910, in latitude 51° North and longitude 23° West during *Lusitania*'s Crossing No. 71 West, the pattern of a rogue wave seems to fit perfectly.

Only thirty-three hours out of Queenstown Harbour, the Cunarder was picking her way through foul weather at 14 knots. At 6 p.m., Captain Turner left the Bridge to go down to dinner, leaving Chief Officer McNeil in command. This

32 *The Scientific American Handbook of Travel*, p.191.

Lusitania punches her way through an enormous rogue wave, proving her great strength in spite of the most appalling conditions imaginable on the North Atlantic. (HFX Studios)

was Turner's last round-trip voyage aboard *Lusitania* before transferring to *Mauretania*. When he left, the wind was out of the west – dead ahead – and the ship's bow was ploughing through one mighty wave after another; Quartermaster Riddey was at the wheel within the Wheelhouse, Relief Quartermaster Harding was standing at his side, and Bridge Boy Tommy Hughes was also on the Bridge. Third Officer Storey decided to brave the elements to climb up on the Bridge roof to read the compass there. No doubt all were relieved when the gangway door slid closed behind him, restoring a measure of calm to the Bridge.

Everything seemed foul but typical, but in a heartbeat everything changed. First, the liner shipped a particularly solid swell, and her bow dropped down into the trough behind it. McNeil, already holding on for dear life, must have been horrified to see what was on the other side of the watery valley: an enormous 'accumulative' wave. It was so mammoth that it shut out the sky and horizon from his vantage point, normally 65ft above the water; outside, Storey had made it up onto the roof when he, too, spotted the wave coming. There was just enough time for the officers to see the wave's approach, but only a couple of seconds to react. McNeil rushed for the Wheelhouse door to give an order to Quartermaster Riddey; outside, Storey realised he could not get back inside before the ship was inundated, and threw himself flat against the deck, locking himself around a support for the compass platform; young Tommy Hughes had the same instinct, even though he was inside the Bridge. In that moment, the officers and men could do nothing but hang on for the ride and pray their ship survived what was coming.

Before *Lusitania* could rise up onto the monstrous sea, it broke above her and collapsed squarely onto her Forecastle and forward superstructure. The whole bow of the liner was pounded underneath the ocean, and solid water engulfed the Bridge before running down the length of the ship moving aft. In a matter of only a couple of seconds, the liner's orderly control centre devolved into waterlogged chaos. The impact smashed in the Bridge's steel face until it was flush with the row of telegraphs and instruments; it pulverised the windows overlooking the bow, also smashing their teak reinforcing screens into kindling. Suddenly, the wave was inside the Bridge as well – the water was 3ft deep and moving fast. McNeil was picked up by the torrent and tossed into the Wheelhouse through the door. In that shocking moment, Quartermaster Riddey gripped the ship's wheel tightly, probably feeling that it was the least likely object in the area to be moved by the rushing water; he was no doubt quite surprised when the wheel was wrenched right off of the telemotor, and they were both thrown into the Chart Room wall. Relief Quartermaster Harding was tossed like a ragdoll, bruising his leg against a bulkhead. The water shorted out the electric lights, adding to the mayhem.

Chief Officer McNeil struggled to his feet, pulling himself through the Wheelhouse door and back out onto the Bridge. The scene must have been astounding, with 3ft of water washing about, and the raging wind outside howling in through shattered and disfigured windows. McNeil spotted a white-gloved hand in the water that was swirling across the room. It belonged to Bridge Boy Hughes, who was just about to be swept out with the current. At the last moment, his hand found a support stanchion and he clutched it, saving himself from nearly certain demise. The force of the water within the Bridge was strong enough to embed a large fragment of teak wood 2in into the wooden case of the fire detection gear. From the Bridge and Wheelhouse, the flow surged aft into the Officers' Smoke Room and the Officers' Quarters, inundating them, as well.

Outside, as the ship's bow rose back up out of the sea, Third Officer Storey must have been quite surprised to find he was still alive; as the remnants of the wave washed astern, it did an astounding amount of damage. The same short that had plunged the interior of the Bridge into darkness also knocked out the electrical masthead and sidelights. The starboard companionway from the Boat Deck up to the Bridge was torn right off; its counterpart on the port side was nearly torn away, and eventually had to be lashed in place. Boats Nos. 1 and 3 were smashed into the deck and damaged beyond repair, their enormous davits badly twisted. What little was left of Boat No. 1 swung in and almost completely blocked the Boat Deck.

With no wheel on the Bridge to control the ship's direction, a new problem presented itself, as her bow began to drift to port until she was pointing south, presenting her broadside to the sea. The engines were stopped until a measure of control could be regained. In the bedarkened Bridge, the water may have been draining away, but the mess left behind was astonishing; the sight of the dismounted ship's wheel and the crumpled, dazed form of Quartermaster Riddey in the corner of the Wheelhouse would, under other circumstance, have been humorous.

There was no time to laugh, however – the situation had to be brought under control quickly. Orders were given to get the aft steering station to control the rudder; meanwhile, Chief Officer McNeil and his men began to work on the lights and on reattaching the wheel. After ten minutes, the wheel was back in place, and another five minutes passed before the lights were turned back on. It was only once these tasks had been completed that McNeil had a chance to get a good look at himself; he noticed that his uniform coat and shirt were not only sopping wet, but they were also bloodstained. He looked for the source of the blood, and found that his forehead had a long gash across it, while his chin was also cut; in the rush and confusion of those fifteen minutes, he had never felt the injuries. Forty minutes after *Lusitania* had survived her encounter with the rogue wave, the ship was again under way and in control, moving at a cautious 10 knots.

A thorough inspection was done, and the Bridge was cleaned up. Captain Turner was no doubt astonished at his return to what had been, only a few minutes before, the organised nerve centre of his vessel; what glances and words were exchanged between him and his Chief Officer can only be surmised. As clean-up continued, it was discovered that a good many items from the Chart Room had been washed out to sea, including the ship's log. The good news was that the liner was not taking on any water from burst hull seams; however, there was some disturbing evidence within her frame of the wave's strength. Bulkheads in crew spaces forward were bent, there were broken doors, and the copper pipes connecting the hoisting gear were bent out of shape.

By noon Tuesday, the battle-scarred liner had logged only 319 nautical miles, and the weather continued foul. Ahead of her, on Saturday morning, the Nantucket Lightship broke loose from its moorings and it looked as if she and her nineteen crewmen could be swallowed up by the sea. The lightship's crew kept in constant contact with the Newport wireless station; the revenue cutter *Acushnet*, located near Woods Hole, Massachusetts, was even put on notice to be ready to respond should she send a call for assistance. Despite the Atlantic's foul temper, *Lusitania* arrived at Sandy Hook on Friday morning; the weather was still too rough to enter New York Harbour, however, and she was forced to anchor until Saturday morning. Once she tied up at Pier 54, her passengers stepped gratefully ashore, most of them very glad that their ordeal was over and that they had found *terra firma* once more. One of the First Class passengers, the Honourable George Keppel, told incredulous reporters: 'I enjoyed the trip, although we had big seas.'[33]

The return trip to Liverpool between Wednesday, 20 January and Tuesday, 26 January was also miserable. It started off all right, but on Sunday the ship steamed straight into a northerly gale. At the peak of the storm, a First Class passenger was tossed to the floor of the Dining Saloon, injuring his back, and a woman in Third Class 'temporarily lost her reason', in the delicate vernacular of the time. The ship's call at Fishguard was cancelled and she arrived in Liverpool a day late, although safe. It had certainly been a memorable final round trip for Captain Turner as *Lusitania*'s Commander.

On Crossing No. 101 West, during February 1911, Captain Charles brought the ship though another series of astoundingly rough gales that came from every point on the compass. The wind speeds, reportedly up to 90mph, matched those of a strong Category 1 hurricane; it blew so hard that the paint along the ship's sides was ripped right off the steel plating. Some of the waves were again strong enough to nearly engulf the Bridge, but once more, the great ship survived relatively unscathed and landed her passengers safely.

Lusitania made another frightful crossing in late October 1913. Picking her way westward towards New York under the temporary command of Captain Arthur Rostron, she ran into what her log described simply as a 'whole gale; high seas'. With winds of 80mph and tremendous waves, the ship forged bravely forth. One particularly rough sea on the afternoon of Tuesday, 28 October swamped the ship's stern, sending solid water down into the Stewards' Quarters and thoroughly dousing everyone and everything inside. Meanwhile, two Able Seamen – Patrick Daly and George Brown – found themselves assigned the unsavoury task of trying to get the falls of the starboard emergency boat sorted out so that it could be swung back in. Just then, the ship lurched to starboard and shipped a huge sea. Daly saw it coming and leapt onto the deck and broke his leg; Brown held on to the lifeboat for dear life as the wave took it and turned it completely upside down, receiving some bruises in the process. Despite these challenges, the liner averaged 23.74 knots on the crossing.

Through all of this, the two Cunard speed queens were clearly very strong and seaworthy liners. In February 1913, *Mauretania* actually found herself poised with her bow

33 *The New York Times*, 16 January 1910.

and stern sitting on two separate mountainous seas, with her midships area sagging unsupported between. Yet this scenario, for which ship designers often calculated stresses but which was hardly ever seen in reality, left *Mauretania* without a single rivet slack in her hull – a very pleasing performance.

The two sisters may have been strong and safe in rough weather, but they were not always comfortable. With their ultra-fine bow lines and minimal ascending bow flare, they tended to cut more directly through large waves than other liners, which to a larger degree rose up and over waves. This meant that solid water was often shipped over the forward decks in bad weather, often right up to or over the Bridge before running aft down the Boat and Promenade Decks. This produced a harsher ride for the ships' occupants than they might have wished for, but this fineness of hull was necessary to their maintaining high speed; it was simply a feature of their design, never posing any real threat to their safety.

Sometimes poor weather could lead to other risks. Daunt's Rock, 4 miles off the coast of County Cork, Ireland, was not only the official start and end points of the transatlantic crossing, but it was also a known hazard to navigation, despite the presence of a lightship close by. The Cunarder *Ivernia*, picking her way towards Queenstown from Boston in heavy fog on 24 May 1911, ran afoul of Daunt's Rock. The ship's starboard bow was damaged badly, and she limped into Queenstown Harbour to be beached. Temporary repairs were carried out, and *Ivernia* later returned to service, but the whole affair prompted discussion about safety signals around obstacles like Daunt's Rock.

As the press discussed the subject, Captain Turner told a reporter from New York's *The Outlook* about a recent hair-raising trip he had made on *Lusitania*. Throughout the eastbound crossing, the weather had been so foul that he had been unable to make a single celestial observation. Running on dead reckoning across the Atlantic, the time eventually came when he knew, almost as if by some sort of sixth sense honed by years of experience, that he was not far from running right into the Irish coast. Captain Turner was widely acknowledged to be 'one of the most

This extremely rare, one-of-a-kind photograph came from an amateur stereoview card. The picture was taken on 23 April 1908, as *Lusitania* was coaling in Liverpool. It would appear that the black paint along her starboard hull had recently been retouched. (Authors' Collection)

skilful navigators afloat', yet he had to use every available tool at his disposal to bring the great liner through safely. 'Soundings, fore and aft, were of no avail, so judging the ship to be near Kinsale he headed the huge liner round, went dead slow, and pointed for the shore,' no doubt praying that his first sight of land wouldn't be when he grounded on it.

Soon enough, he heard two gun signals sound in quick succession to starboard. 'He threw the helm over to port, and told the officer on the bridge to time off six minutes, when again the two guns were heard, and that meant the Head of Kinsale was close at hand, and, twelve miles further on, Daunt's Rock.' *Lusitania* had been saved through her Captain's great caution on that occasion, but navigating in poor weather was more dangerous than most of her passengers generally thought. The article concluded, 'The problem of safety at sea … is only half solved by that which is inventive and automatic. It still requires a man who makes no error of judgment, who is cool, calm, collected, and who knows the seas as if they were the streets of a mapped-out city.' Captain Turner obviously had the experience and skills required to cope with any imaginable danger that the sea could throw her way.[34]

Storms, fog and close calls in escaping the sea's wrath may have generated oft-told tales, but perhaps the most important part of life aboard *Lusitania* was, for her officers and crew, a sense of order and routine. After the high-stress and excitement of the maiden voyage, life for the crew began to settle down quickly. What had once seemed enormous and overwhelming soon became familiar and mundane. Many crew members signed on for numerous voyages, for months or years at a stretch. Officers were assigned by the Cunard Company, and they also often stayed for lengthy tenures. Crew members in various departments found *Lusitania* was a good job, and she became much beloved by those who manned her. Trip after trip found many of the same crew members working together. This created, in a very real sense, a home and family atmosphere aboard the huge ship. In turn, this 'family' created an overall personality for *Lusitania*. No doubt, to some extent, the crew's good rapport and positive sentiments towards their vessel had something to do with the fact that *Lusitania* was later remembered as a 'happy' ship. Other vessels over the years suffered from the reverse, and it showed. This sentiment, be it positive

34 *Poverty Bay Herald*, 7 June 1911, p.7.

Above: 20 November 1908: *Lusitania* has tied up at the completed Pier 56. Pier 54 has not been built yet, and the interconnecting structure between the two is still in the early phases of steelwork, through which *Lusitania*'s name can be seen. The new structures were designed by Warren & Wetmore, the company that would go on to design the Grand Central Terminal Building. (Library of Congress, Prints & Photographs Division, Authors' Collection)

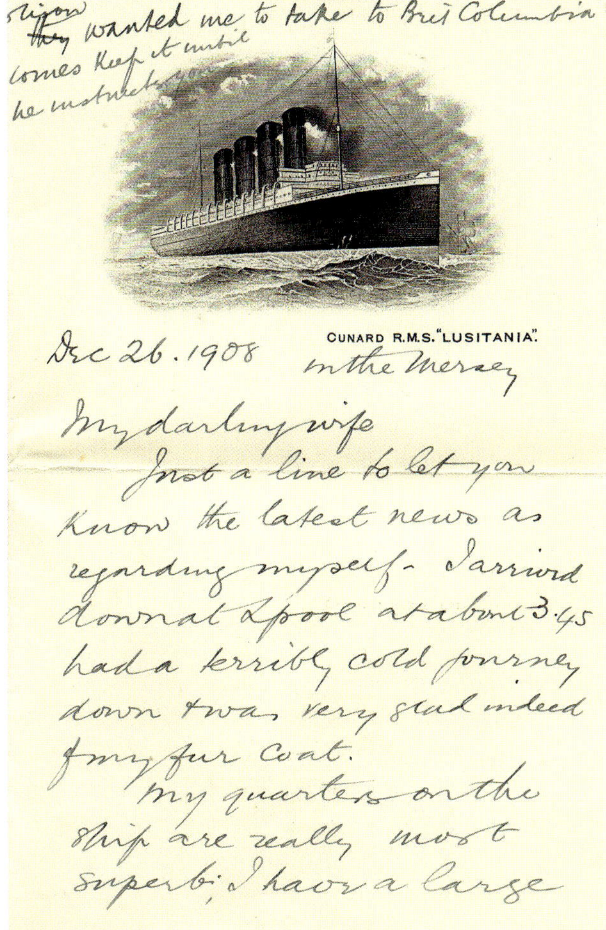

[illegible] ~~they~~ wanted me to take to Brit Columbia
comes Keep it until
he instructs you

CUNARD R.M.S. "LUSITANIA".

Dec 26. 1908
in the Mersey

My darling wife

Just a line to let you know the latest news as regarding myself - I arrived down at L'pool at about 3.45 had a terribly cold journey down & was very glad indeed of my fur coat.

My quarters on the ship are really most superb; I have a large

Left: This letter, written aboard *Lusitania* on 26 December 1908, was penned by First Class passenger Austin Partner. It was addressed to his wife. Just three and a half years later, Partner would perish in the sinking of White Star's *Titanic*. (Authors' Collection)

Lusitania's prow is an imposing sight at her New York pier in this photo, likely taken around 1910. (Authors' Collection)

or negative, was just as real and tangible to passengers as if it were a physical object, and it remains just as true today as it was then.

This 'family' started at the top, with the Captain. Naturally, the senior officers also played a part, but the Captain was the true driving force, the 'father figure' aboard. The men who were appointed as 'Masters of the Atlantic liners' during that era were an interesting breed of individual; in *Lusitania*'s day, most were still from the ranks of sailing ships, a select group that grew ever smaller with each passing year. Most had colourful personalities. Some loved spending time with passengers, telling tales and associating with society's elite, while others viewed this social intercourse with indifference or contempt. Some were warm and personable, while others were gruff and short-tempered, especially towards officers and crewmen. Some were strict disciplinarians, driving their ship and crew to operate at peak efficiency; others kept a somewhat easier eye on efficiency. The crew viewed some Captains with cool distaste, while others inspired their unswerving dedication and loyalty. Passengers also grew attached to certain Masters, some even following favourite skippers from ship to ship. Once disconnected from the umbilical of company management ashore, each Captain could run his ship with a certain personal flourish; whether they were loved or hated by passengers and crew, all of them were skilled seamen. They were also alike in that many seemed filled with a certain wanderlust in life, finding a home in their vessels until the sea had become their natural element.

Lusitania had several different commanding officers during the seven and a half years she was in service. Her first was Commodore James B. Watt. When he took command, Watt was 65 years old – two full years past Cunard's normal retirement age. However, special exemption had been granted to the Scottish Commodore so that he would be able to take the great new Scottish-built liner out and break her in. Watt had started under sail, joining Cunard in 1873 and rising through the ranks with distinction. Most of the previous crack Cunard ships, like *Lucania*, *Campania* and *Carmania*, had all come under his captaincy in preceding years. As *Lusitania*'s first skipper, Watt set the tone for the ship's entire career; it was under his command that the liner and her crew truly began to settle into a routine of service. For nearly a year, Watt held the great ship's reins.

By late summer 1908, however, Watt's retirement was imminent; with his departure from active service, *Mauretania*'s Captain John Pritchard would be left with seniority and would become the Commodore of the line. The decision was made to have Watt's position aboard *Lusitania* filled by Captain William Thomas Turner, then Master of *Caronia*. Watt's last round-trip voyage, the liner's seventeenth, began in late October 1908. On the westbound crossing, the ship encountered strong seas; at one point, the liner buried herself so deeply in the sea that the wave swept down the decks and found its way into the First Class Grand Entrance via an open door. Seawater cascaded down the elevator shaft, down the Grand Staircase, and thoroughly drenched some of the passengers as it went. But to Watt, the crossing was no doubt memorable for another reason: his daughter Helen was travelling to the United States for her wedding. Once the ship landed in New York, she would be married to her fourth cousin, a man named Gordon Watt from Chicago.

In the early morning of 30 October 1908, Captain Watt stood on *Lusitania*'s Bridge as she steamed slowly up the Upper Bay under the eye of the harbour pilot. Watt's mind was no doubt churning; his face was noticeably stern. As the skyline of New York came into view through the mists, filled with tall buildings of every sort, it must have been a truly bittersweet moment. This was the last time he would greet Manhattan, a city that had become a home away from home of sorts. The stay in New York and the return crossing to England went smoothly; once *Lusitania* returned to Liverpool, Captain Watt turned command over to Captain Turner. In all, Watt had commanded the ship for one year and two months.

Captain William Turner leaning on a ship's telegraph. Turner had been chosen by Cunard to replace Captain Watt when his tenure of command ended. Turner remained in command for over a year before moving on to captain *Mauretania*. (Authors' Collection)

Captain Turner was a remarkable officer; under his leadership, *Lusitania* would attain some of her most remarkable accomplishments. Born in October 1856, Turner was 51 years old when he took command. He had gone to sea aged 13, working under sail for some time; although he was quite proud of having served aboard sailing vessels, he did not particularly relish his experiences during that period, recalling harsh words from superiors, weevil-infested biscuits, and severe punishments. His goal had always been to command a ship, and in 1878 he had taken a major step forward towards attaining this dream when he joined Cunard as *Cherbourg*'s Third Officer. With his career going in the right direction, he was married in August 1883. Unfortunately, within a few years, he discovered that Cunard would not appoint an officer to command one of their ships unless he had first commanded a vessel of another line. So Turner returned to sail for a short spell in 1889, and was appointed Master of the *Star of the East*. After one round-trip voyage to Australia, he returned to Cunard and continued moving up the ranks, serving with distinction during the Boer conflict as *Umbria*'s Chief Officer and even earning a medal for his work. In 1883, he had saved a young boy from drowning in Liverpool, plunging into icy water to make the rescue; this service had earned him a second medal, from the Royal Humane Society of Liverpool.

His first Cunard command was *Aleppo*, and in short order, he moved through *Carpathia*, *Ivernia* and *Caronia*. Achieving command of *Lusitania*, one of the two largest and fastest vessels in the world, must have seemed like a remarkable accomplishment to him, but he had certainly proven his worth to Cunard. Turner was a skilled seaman, never having experienced a major emergency while in command. It was noted, however, that he had a somewhat gruff manner; in particular, he had little patience for First Class passengers, and apparently attempted to slip out of his social obligations whenever possible. He was described as a man of medium height, with a bright, if weather-beaten face; he sported a distinctive row of crow's feet around his eyes when he smiled. His niece, 14-year-old – as of summer 1909 – Mercedes Desmore, lived in New York, and it was a favourite pastime for him to spend time with her while he was there. Perhaps he lavished this attention on her because his own family situation was not going particularly well. His wife had moved out of their home in 1903, taking their two sons with her. With things going poorly on the home front, at least Turner could take comfort in his career, being in charge of the mighty *Lusitania* – and better things were ahead for him.

In December 1909, Cunard decided that no man over the age of 60 would henceforth command *Lusitania* or *Mauretania*. This policy alteration meant that *Mauretania*'s Captain Pritchard, who was over 60, would be retiring almost immediately. His departure meant that Captain Robert C. Warr became Cunard's new Commodore. However, because Warr was 61 years old, he was placed in command of *Umbria* instead of one of the two crack sisters. Captain Turner was selected as *Mauretania*'s next skipper – primarily because of his fine performance as *Lusitania*'s Master – and it was decided to bring 45-year-old Captain James Charles aboard *Lusitania* in his place. The change came swiftly. At the conclusion of *Lusitania*'s Voyage No. 36, on 26 January 1910, Turner just barely had time to join *Mauretania* on her next westbound passage.

It was another month before *Lusitania* departed on her next crossing, but when she left Liverpool on 26 February Captain James T.W. Charles was resolutely in charge. Not only was he the youngest man who would ever command *Lusitania*, but he would also hold that position longer than any other Captain during the ship's career. Charles was rather tall, standing almost a full head higher than most of his officers. He stands out as a particularly likeable Captain. An example of his pleasant manner came during his first westbound crossing in command, when a 19-year-old stowaway was discovered aboard. He had boarded in Liverpool with the Third Class passengers, and had not been discovered until two days after the ship left Queenstown. When he told Captain Charles his story, it was obvious that Charles immediately took to the young man, commending him for his 'supreme nerve', and even giving him a gift of some cigarettes when he met him on deck later. When the liner landed in New York, the young man was let go without prosecution.

In 1911, while in charge of *Lusitania*, Captain Charles received the distinction of being appointed Sir James Charles. While he was taking *Lusitania* out of New York as the new White Star giant *Olympic* was arriving on her maiden voyage, Charles failed to give the incoming ship a customary salute; some felt that this was an intentional snub, but others more charitably speculated that he was busy both with taking his ship out of dock and with absorbing the news of his new title. Captain Charles grew quite attached to the ships that he commanded; in 1928, then-Commodore Charles was on his final voyage in command of *Aquitania*, and he passed away before the trip was over. Many believed that he had quite literally died of a broken heart. While he was in charge of *Lusitania*, Captain Charles did something rather unique, making an endorsement for the Auto Strop Safety Razor; it was quickly run as a full-page ad in magazines, carrying his photograph in Cunard uniform. Captain Charles guided the ship on forty round-trip voyages, no fewer than eighty crossings, ending in October 1912. These proved some of the most turbulent months of *Lusitania*'s career before the Great War. There was a certain loss of prestige with *Olympic*'s introduction into service, and there were months after *Titanic*'s loss when big ships like *Lusitania* were viewed with some scepticism even after they were provided with enough lifeboats for everyone aboard.

Through it all, however, Captain Charles watched over the ship, her passengers and crew, and the liner continued her successful and safe service. When *Lusitania* went in for work on her turbine engines in October 1912, she was out of commission for nearly two months, and change was in the air. Captain Charles was transferred and replaced by Captain Daniel Dow. The ship returned to service under Dow's care, departing Liverpool on 13 December 1912. On this crossing, she was performing the annual Cunard Christmas mail run in *Mauretania*'s stead. Captain Dow's command of *Lusitania* lasted only one round-trip voyage, however, for when she returned to Liverpool on 31 December, she was laid up again, once more due to trouble with her turbine engines. When she re-entered service on 23 August 1913, she was back under the command of Charles. This time he was only to stay in command for two months, or three round-trip voyages. At the conclusion of this stint, Captain Turner was appointed to become Master of the new *Aquitania*, Captain Charles was moved to *Mauretania* in his place, and a gap opened up on *Lusitania*'s Bridge once again.

On 25 October 1913, *Lusitania* started her next round-trip voyage, No. 81, with Captain Arthur H. Rostron, in charge. Rostron was the famous skipper of *Carpathia* who had rescued *Titanic*'s survivors a year and a half before. Rostron's command of the liner was only eighteen days in length, and included the foul-weather trip referred to previously. This was the shortest term that any of *Lusitania*'s Masters would put in. After Rostron's single round-trip voyage, Captain Dow again took *Lusitania*, starting with Crossing No. 163 West, which began on 22 November.

Daniel Dow – a Scotsman by blood, but born and raised in County Cork, Ireland – was well known all over the Atlantic as 'Paddy' Dow. He had acquired the nickname from the Liverpool Irish stokers who were acquainted with him. His other nickname was 'Fairweather' Dow because of his apparent tendency towards seasickness in all but the fairest of weather. Dow was well known as a skilled and cautious skipper, one who could always be relied on to make correct decisions in an emergency situation. He had served with Cunard since he was 28 years old; at that time, he had already been at sea for a dozen years. In the winter of 1914, while commanding *Lusitania*, Dow successfully co-ordinated the rescue of crew members from *Mayflower*. For their efforts in this crisis, he and his officers received an

A razor advertisement featuring Captain James Charles while he was in command of *Lusitania*. (Authors' Collection)

Captain Arthur H. Rostron, who had narrowly missed a chance to be *Lusitania*'s Chief Officer on her maiden voyage when he was promoted to his first command, was a temporary skipper of *Lusitania*. He would go on to command *Mauretania* in later years. (Library of Congress, Prints & Photographs Division, Authors' Collection)

Captain Daniel Dow, a Scotsman by birth, but raised in Ireland, loved to regale passengers with humorous tales, and used his pipe as a prop when retelling them in his 'racy Hibernian style'. (Authors' Collection)

official presentation at the Town Hall in Liverpool. He was described as rugged in appearance, and was known to run a tight ship. He was also noted to be quite sensitive and solicitous, a 'genial gentleman' that simply refused to be flustered no matter what the provocation. He was also strong-willed and was backed up by an 'inexhaustible fund of shrewdness'.

Unlike Will Turner, Dow relished socialising with his passengers. He was recalled as brimming with 'lively anecdote', which he could tell in a 'racy Hibernian style'. Two of his frequent passengers were Olga Petrova – an accomplished Broadway and film actress and storywriter – and her husband, Dr John Stewart. In her memoirs, published nearly thirty years after the event, she recalled Captain Dow as a 'brilliant and tireless storyteller and a mine of information on the lives of the great men of the sea'. She recalled that his 'special hero was Lord Nelson and he would, if unchecked, be quite willing to sit up till dawn reciting his exploits', using his beloved pipe as a prop.[35]

Around Christmas 1912, Captain Dow recalled rather nostalgically: 'I passed my first Christmas at sea off Cape Horn in a snowstorm as apprentice on a 500-ton bark, and this will be my twenty-first at sea. The *Lusitania* could carry my old clipper packet on her after deck, but for all that the old days were happy days. The plum duff tasted far better to me without any music or frills, except a few chanteys from the star performers in the foc'sle and half deck, than it does these days, when liners have been transformed into floating hotels.' Captain Dow remained in command until the conclusion of her 198th crossing. His term as Commander lasted fifteen and a half months, taking her right through the opening months of the Great War.

Lusitania's shipboard 'family' also included her senior officers. Cunard's officer hierarchy was not exactly identical to the better-known hierarchy of the White Star Line. White Star placed immediately under the Captain a Chief Officer, and he was succeeded by six officers numbered First to Sixth in descending rank – eight men, in total. Cunard's chain of command bore the same number of officers, eight all told, but was arranged differently: immediately under the Captain was the Chief Officer, and he was succeeded by Senior and Junior First, Senior and Junior Second, and Senior and Junior Third Officers; a ranking understandably confusing to the uninitiated. Each of these Executive Officers had

35 *Butter With My Bread: The Memoirs of Olga Petrova*, Mme Petrova (Bobbs Merrill, Indianapolis: 1942), pp.245–6.

to be equipped with Master's Certificates, and some had more than one to their credit.

Each of these men had specific duties to attend to at various times during the ship's routine. All but the Senior Second Officer were assigned specific watches to maintain during the course of the day. The first watch of each day began after sundown, at 8 p.m.,[36] and they ran their course in this way:

Watch	Time period	Officers
First	8 p.m. –12 a.m.	Junior First/Junior Third
Middle	12 a.m.–4 a.m.	Senior First/ Senior Third
Morning	4 a.m. –8 a.m.	Chief Officer/ Junior Second
Forenoon	8 a.m.–12 p.m.	Junior First/Junior Third
Afternoon	12 p.m.–4 p.m.	Senior First/ Senior Third
Dog watches	4 p.m.–6 p.m.	Chief Officer/ Junior Second
	6 p.m.–8 p.m.	

The Senior Second Officer was responsible for mail cargo; however, if another officer were unable to maintain his watch, the Senior Second would do so in his place. Of the two officers who were on watch at any given time, the senior of the two men was designated the Officer of the Watch. As such, everything that happened during his duty shift was his responsibility. If something unusual or extraordinary occurred, it was also his responsibility to inform the Captain on the matter. The Captain himself did not hold a set watch, but was on the Bridge quite often throughout the day, keeping a close eye on the goings-on. Additionally,

Looking down on the Bridge and forward Boat Deck from the Crow's Nest. The view clearly shows the extra compass platform installed atop the Bridge; the frame for the canvas sun shades has been extended to five cross-braces, but not six as seen around 1911; the lack of post-*Titanic* lifeboats amidships indicates the photo could have been taken in 1910. Two officers stand on the starboard Sun Deck, just abaft a ventilator. (Stuart Williamson Collection)

in cases of fog or emergency situations, the Captain was expected to be on the Bridge until the danger had passed – a duty that, in sailing on the often foggy, frequently hazardous North Atlantic, could prove to be a mind-numbing marathon.

While *Lusitania* was in port, there were watches to keep as well; three senior officers kept a standard 9 a.m. to 5 p.m. duty shift, although this could be either shorter or longer depending on the situation. The junior of each pair of men who held at-sea watches together maintained that same watch on their own while in port. This allowed them to handle any situation that developed day or night and take appropriate action, and it also gave them valuable

36 According to numerous sources, including *Mauretania: Landfalls and Departures of Twenty Five Years*, Humfrey Jordan, Patrick Stephens, Wellingborough: 1988, and *Outward Bound*, by John Gould (1905/1913). Another period reference, *The Scientific American Handbook of Travel*, p.102, notes that the first watch of each day began at noon. In either case, the watch breakdown was the same.

experience. While passengers were embarking or disembarking the ship, there was another set of duties to tend to. The Junior Third Officer was charged with the forward gangways and luggage; the Senior Third with the after gangways and baggage. The Junior Second was concerned with gangways and specie; the Senior Second also looked after specific gangways while keeping an eye on the mails, which also happened to be his primary at-sea responsibility. The Junior First Officer was stationed on the Bridge, while the Senior First Officer took charge of all the seamen. The Chief Officer held in his hands the reins for the entire embarking or disembarking operation.

Lastly, while entering or departing port, there was yet another round of duties to be handled. The officers were paired off with the same partner they kept watch with. The Chief and Junior Second Officers were stationed on the Forecastle head. The Senior First and Third Officers took up position on the Stern Bridge and the after mooring station. The Junior First and Third Officers were positioned on the Bridge. The Senior Second escorted the pilot to and from the Bridge, and also had to keep an eye on the ship's shell doors, any tenders that were servicing the ship, and he also had charge of the gangways and mails.

All these duties constituted routine shipboard operations; this was by far the way most time passed aboard, whether in port or at sea. However, there were also rare occasions when emergencies arose; at such times each officer also had a specific duty assignment. These could run the gamut from controlling fire extinguishing systems to rigging collision mats. Emergency situations were drilled frequently, and each officer knew that he was responsible for carrying out his assigned duties, should the need arise, in a timely manner. In such situations, life and death – either for individuals or in dire cases for everyone on the ship – quite literally came into their hands. This was a fact that these men knew only too well; no doubt each hoped every day that such a situation would not arise, but also doubtless hoped that if one did they would be able to discharge their responsibilities well.

In addition to their regular watches, emergency preparedness and navigation duties – which included not only taking regular fixes, but also keeping the ship's charts up to date – *Lusitania*'s officers were heavyweight paper-pushers. No fewer than half a dozen various ship's logs had to be maintained regularly; manifests had to be signed off on; reports had to be filled out and filed correctly. The list must have seemed endless. The Captain was also busy; he kept a customary set of morning rounds, inspecting the ship from stem to stern with his senior officers before receiving reports from the various department heads. By the end of each day, the officers were exhausted; a junior officer from the White Star Line once put it this way: 'You must remember that we do not have any too much sleep and therefore when we sleep we die.'[37]

All of this work left little time for socialising with society's elite. Although there was a specific table set aside for the Captain and his officers in the First Class Dining Saloon, for obvious reasons it was never fully populated. When officers encountered passengers, they were to be courteous, informative and agreeable. However, their responsibilities were simply too numerous to allow them to whittle away time in social niceties beyond those that were absolutely necessary. However, most officers – from the Junior Third to the Captain – enjoyed a certain 'he's in the club' sort of treatment from many First Class passengers who travelled with them. This was despite the fact that a majority of them had worked their way up from extremely humble origins. Anything that they said was taken by even the wealthiest or most influential of passengers with extra consideration and thought; it must have seemed quite ironic to the officers that they were granted such a high status by their rich and famous passengers. It must also have appealed to their vanity at times, and there were no doubt occasions that they thoroughly enjoyed the at-sea hierarchy.

Although these men frequently had girlfriends, fiancées or wives on land, all of them were young, no doubt frequently cursing their inability to linger and chat with some of the more ravishing ladies who were in their official, if detached, care. Conversation would frequently have turned to discussing these lovely ladies and commiseration over not being able to make more leisurely contact with them. There would also have been some gossip and head-wagging over the often absurd incidents that took place on each voyage ... 'The gamblers took how much from so-and-so?' 'You mean that there was an argument over who had made the first bid on that number in the ship's pool?' ... and the like. The officers discussed these goings-on in the Officers' Wardroom, around their meal table, and even on the Bridge during quiet moments. More specifically, quiet moments when the

37 Testimony of Fifth Officer Harold Lowe to the US Senate Inquiry into *Titanic*'s loss.

These two photographs were dated to summer 1912, while *Lusitania* was in New York.

Far Left: A pair of officers enjoy a rare moment of shipboard socialisation with two women. Questions abound, though: were the women passengers? Or were they visiting the officers while the ship was in port? Were they romantic interests, or family or friends? Sadly, no back story has survived. This photo was taken on the Stern Docking Bridge. (Authors' Collection)

Left: This photo shows some interesting details: first, the 'bus stop'-style deck benches seen in the Second Class Boat Deck area in earlier years are now seen in this photo, yet did not remain there for long. Additionally, the quartet sit atop a skylight that was not there when *Lusitania* was first built, one that looks rather like *Mauretania*'s. Yet another photo from this period confirms that these skylights were added to *Lusitania* and remained there until 1915. (Authors' Collection)

Captain wasn't around, because the Bridge was expected to be as quiet as a crypt unless conversation pertained to navigation. During their voyages together, these officers and men frequently formed solid friendships that lasted even when they were reassigned to other vessels.

On 29 May 1912, the Cunard Company's Board of Directors met and decided to add another member to this family of officers. They created a new position, namely that of the 'Staff Captain'. About a week after the decision was made, the Directors outlined that the Staff Captain was responsible for ensuring passengers' safety, primarily by keeping track of lifeboat drills and the general operation of the ship; the new position was doubtless created in direct response to the *Titanic* disaster of only six weeks before. 'It will make a good deal of difference to the captain and will help the Chief Officer,' Charles Sumner explained in New York.

The men appointed to this position were fully-fledged Masters of other vessels, not ordinary Executive Officers, and would only serve aboard *Lusitania*, *Mauretania* and *Aquitania*, once she entered service. Smaller Cunarders, like *Caronia*, would not be endowed with a Staff Captain, but would instead have an extra Executive Officer added.

Captain Samuel 'Sandy' G.S. McNeil (RD, RNR), formerly *Lusitania*'s Chief Officer but now serving as Master of *Albania*, received a letter from Cunard dated 30 May informing him that he would become *Mauretania*'s Staff Captain on her next crossing, due to begin on 1 June; this short notice certainly made for a hasty transfer. *Veria*'s

Pilot Collins about to climb the stairs leading to the Bridge, probably around 1913. Although harbour pilots were not official members of the ship's crew, their presence as she entered or departed port was a constant of *Lusitania*'s routine. Notice the half-raised privacy screen through the open window; between the steps is a brass plaque that says 'NAVIGATORS ONLY'. (Library of Congress, Prints & Photographs Division, Authors' Collection)

Captain Jerry F. Simpson received an almost identical letter on 6 June, informing him of his appointment as *Lusitania*'s Staff Captain beginning on her next crossing, which would start on 8 June. The letters clearly pointed out that the Captain was still to be 'in supreme command', and that they were to consult him in any required matters so that there could never be a conflict of interests. In consolation for their transfers, which might have seemed like a demotion of sorts, both men were given a raise in pay of £25 per year, for a full £300 annual wage. Attached to each letter the men received was a two-page list of some twenty-three separate duties that came with the position. These included the general discipline of the crew, general cleanliness of the ship, taking charge of lifeboat drills, handling passenger complaints, seeing to the ventilation of the ship, making the 10 a.m. rounds with the Captain, making unscheduled visits to department heads throughout the day, and presiding over a table other than the Captain's in the First Class Dining Saloon. The Staff Captain's normal day would run from 7 a.m. to 9 p.m., after which time he would make his daily rounds through the ship; he would not be required to stand a four-hour watch, unless one of the other officers could not fulfil theirs.

Looking back, Cunard was clearly taking some of the social obligations and navigational details off the hands of the Masters of each of their three largest ships. This would allow them to focus more on safe navigation; the Staff Captain's duties involving lifeboat drills show just how seriously Cunard was taking this sort of thing in *Titanic*'s wake. Furthermore, this decision was a wonderful public relations move; just knowing that Cunard was concerned enough to appoint an extra officer primarily to look after lifeboat drills reassured nervous passengers about their safety at sea. Both Staff Captain McNeil and Staff Captain Simpson wore the same uniforms as Captain Turner and Captain Charles, and they were given accommodation on A Deck forward, near the Captain's Suite, which had previously been reserved for First Class passengers. When *Mauretania* arrived in New York on 7 June 1912, Staff Captain McNeil showed that he was taking his responsibilities seriously by conducting a full lifeboat drill, lowering all of these craft to the water.[38]

Although it was the senior officers who were most sought after by the passengers, they were far more likely to be able to spend time with the Ship's Surgeon, the Assistant Surgeon, the Purser, or one of his assistants. Although these men were not directly responsible for *Lusitania*'s navigation, they were members of her senior staff who were specifically selected to spend time with passengers during each crossing; they were also closer to the inner sanctum of the officer corps than most of the crew. 'Avoid asking the officers questions about the navigation of the ship; remember that they have had to answer these questions many thousands of times, and eventually this becomes wearisome even to the most good-natured officers,' passengers were advised.[39] Despite such reminders, how many times the Purser or other officers were asked about the minutiae of navigation or technical matters – or for the 'inside scoop' on the next day's run – is a point that begs an incalculable answer.

When the company was fair, spending time with the passengers could be a joy to the Purser and his staff; when the

38 *The New York Times*, 8 June 1912/The Cunard Archives, University of Liverpool Library/*The Daily Gleaner*, 18 June 1912.

39 *The Scientific American Handbook of Travel*, p.100.

Lusitania's crew stand beneath the Bridge of their ship on the Forecastle and forward superstructure. Judging from the features of the ship visible in the photo, it was taken during late 1908 or 1909. Interestingly, four of the eleven Bridge windows have protective covers on them, either because they had been broken or to prevent them from being broken in bad weather. Three other Bridge windows are open and crewmen stare out through two of them. (Authors' Collection)

Another crew photo, this one taken from the Stern Docking Bridge while the crew stands on and around the Second Class Lounge and skylight over the Smoking Room. Judging from the presence of Pier 54 behind the crew, this must be in 1910 or later, as that pier was not finished before that year. Also interesting is that the port-side staircase to the Lounge roof had been removed at some point and was apparently never put back in place for the remainder of the ship's career. (J&C McCutcheon Collection)

Lusitania's Boiler Room staff pose on the Boat Deck in this ultra-rare 1912 photograph. (Stuart Williamson Collection)

company was dim-witted or outright annoying, it could be a curse. Only the best of 'good-natured officers', who also had a flair for the diplomatic, were chosen to officially care for *Lusitania*'s First Class passengers. Her first Chief Purser was Claude Lancaster. Lancaster had joined Cunard in 1879 as Third Officer aboard *Brest*; poor eyesight, however, had taken him from the Navigation Department into the realm of Purser. He stayed with *Lusitania* from the time of her maiden voyage until the conclusion of her seventy-sixth round-trip voyage in December 1912. The last officer to hold this post aboard *Lusitania* was James A. McCubbin.

Below decks, far away from the comforts of passenger accommodation, there was another group of elite men. Whereas the senior officers, the Chief Purser, and his staff all had duties socialising with passengers, it was *Lusitania*'s Engineering Department that socialised with her machinery, ensuring the liner operated at peak efficiency at all times. *Lusitania*'s first Chief Engineer was, appropriately enough, a Scotsman, Alex Duncan; he would later become the Chief Engineer on other prestigious Cunard liners like *Berengaria*. Scotsmen seemed disproportionately, though appropriately, high in number in *Lusitania*'s Scottish-built Engine Room; they knew and loved their ship and her beautiful engines. She, in turn, seemed to give them the respect and performance they expected of her no matter what the weather. The last of *Lusitania*'s Chief Engineers was Archie Bryce, a good friend of Captain Will Turner.

And so *Lusitania* plied the seas for those seven and a half years, covering the same route out and back like clockwork. Deviations were few and far between, excepting those occasions when, due to poor weather, a stop at Queenstown – or later, at Fishguard – had to be cancelled. In the course of this routine, the ship typically left Liverpool on a Saturday, and made New York anywhere from Thursday night to Saturday morning, depending on the weather. The following Wednesday, usually at 10 a.m., she would depart New York on the return trip, arriving at Queenstown on Monday, usually before noon; she would arrive in Liverpool by late Monday or early on Tuesday. That Saturday, some three weeks after she had begun the cycle, she would start the process all over again. Within this overall three-week routine, however, her schedule did change from time to time as Cunard tried new approaches to pleasing their passengers.

As early as July 1909, there was talk of adding Fishguard, Wales, as an eastbound stop on *Lusitania*'s itinerary. It seemed an ideal port for a stop on the eastbound Atlantic route and would allow overland travel to London via the Great Western Railway (GWR) on a journey of about five hours. If passengers wished to stay in Fishguard overnight, they could do so at the GWR Co.'s Fishguard Bay Hotel. Surveyors were sent out to Fishguard to check the harbour and ensure that it could handle the draught of the two largest ships in the world. They found sufficient water, and also that the Bay was well sheltered. It was protected on the east, south and west by headlands and hills several hundred feet high; to the north it was protected by a breakwater.

With the green light given to bringing the sisters into Fishguard, *Mauretania* called there first on 30 August 1909. *Lusitania* made her first call there on Monday, 13 September 1909 at the end of her thirtieth round-trip voyage. Having encountered strong headwinds through the entire crossing, she had only made 24.82 knots, as opposed to the 25.41 managed by *Mauretania* on her previous crossing; she was carrying only 529 passengers – 181 First, 160 Second and 188 Third Class. By Sunday, it was already clear she would not make her scheduled arrival time in the Welsh port, 'shortly before noon', and wireless messages were sent ahead to inform those waiting for her there.

She first stopped at Queenstown, Ireland, on Monday morning; there she landed seven passengers and the Irish mail. Coming aboard were Captain Tinsley of Cunard's Marine Superintendent Office, who would help oversee

Far left: This image appears to have been taken in New York, probably in or after 1910, as *Lusitania* was departing for Europe. Based on this photographic evidence, it seems the stripe of black paint extended along the top of C Deck in that year was apparently extended along the Second Class superstructure sides for at least a short while. However, this strip was soon replaced with white again along Second Class areas. The ends of the Cunard piers, even once the new piers had been built, were open at the ends, and this could be where the photo was taken. (Authors' Collection)

Left: A Japanese man and his son walk on the aft Boat Deck, just outside the Verandah Café, in June 1909. On the deck beneath the arched openings, a track can be seen that had been installed since the ship entered service; it was used to put portable wooden weather screens in place to shelter those using the café. (Library of Congress, Prints & Photographs Division, Authors' Collection)

the stop at Fishguard, and Percy Knight, the Inspector of Customs for Liverpool, England, who would take charge of inspecting passengers' luggage in Fishguard.

Lusitania left Queenstown at 11 a.m., and was spotted 9 miles west of Strumble Head Lighthouse, just outside Fishguard, shortly before 4 p.m. The weather there was less than ideal; strong north winds had created a vicious swell outside the breakwater. Two tenders were dispatched to intercept *Lusitania*: *Pembroke* would take the mail, while *Great Western* would take passengers and their baggage. The plan to use the smaller *Sir Francis Drake*, which had previously serviced *Mauretania*, to offload passengers was called off in favour of the larger steamer because of the weather conditions.

At 4.22 p.m. *Lusitania*'s funnels appeared outside the breakwater, and by 4.27 those on the tender could see her full, spectacular form as she made for the harbour entrance. Instead of crossing the breakwater, *Lusitania* spent some minutes swerving around and taking up a position 'which formed a direct continuation of the breakwater'. At 4.42 p.m., she dropped anchor.

The 'fussy little tug' *St David's* approached *Lusitania* first, and *Lusitania*'s passengers 'cheered and waved hand-kerchiefs'. At 4.43 p.m., *Pembroke* came alongside, filled with

Looking forward along the Sun Deck towards the Verandah Café, this candid photo shows an interesting detail: the jibs for the two cranes are stored pointed inward over the centre of the ship, and they are tied together to help prevent them from swinging around. (Stuart Williamson Collection)

The mail tender *Pembroke* approaches *Lusitania*'s starboard side during the liner's first visit to Fishguard, Wales, on 13 September 1909. (©Amgueddfa Cymru – Museum Wales)

forty 'alert postal men' from Swansea, and the mail transfer began. As *Great Western* approached, it was said that she curtsied 'very graciously' in the swells; she tied up alongside at 4.58 p.m. Meanwhile, *St David's* continued to circle around the flotilla, 'her minute proportions exaggerated by contrast with the Cunard leviathan'. In fifty-four minutes from the moment *Lusitania* dropped anchor, 958 bags of mail and 178 passengers and their luggage were all transferred to the tenders. A total of 120 passengers from First Class disembarked here, leaving only 53 of their compatriots travelling up to Liverpool on the ship. During the passenger transfer, the swell increased, forcing the crew to adjust the moorings a couple of times; still, the disembarkation went quickly. It was recalled:

> In fact, in the eyes of the passengers, the slight pitching of the boat appeared to add zest to the novelty of landing on Welsh soil, and it was with the utmost of good humour that even the oldest of the ladies essayed to cross the broad gangway.

Except for the loss of one gentleman passenger's straw hat, everything went smoothly. At 5.34 p.m., the last passenger was aboard; one minute later, *Great Western* cast off. At 5.36, *Lusitania*'s anchor was already coming out of the water, and she turned to start for Liverpool, where she expected to arrive around midnight.

The mail was on shore, loaded on the postal train, and pulled out from the station by 5.42 p.m. By 6.22 p.m. the first passenger express train left for London, with the second departing nine minutes later. The 120 passengers travelling to London saved about thirteen hours of travel time. Meanwhile, eight Continental passengers made their way to Dover and boarded the cross-Channel boat there a mere eight and a half hours after stepping off *Lusitania*; they were in Paris by Tuesday morning. Truly, this first call at Fishguard had run like a well-oiled, precision machine. Passengers, especially those travelling for business, were very enthused about being able to save so much time; there was speculation that other liners, such as the Canadian Pacific and Allan Line vessels on the Canadian run, would soon follow *Lusitania* and *Mauretania*'s example and begin calling at the Welsh port.[40]

On *Lusitania*'s second stop at Fishguard, on Monday, 4 October 1909, things went even better. She had made better time across the Atlantic, averaging 25.10 knots despite strong easterly gales; she dropped anchor at Fishguard at

40 Information on this stop at Fishguard, including extracted quote, gathered from *Evening Express and Evening Mail*, 13, 14 September 1909; *The Haverfordwest and Milford Haven Telegraph*, 15 September 1909; *The Tenby Observer*, 16 September 1909; *The County Echo*, 16 September 1909.

2.53 p.m., disembarking 214 passengers and 1,060 sacks of mail before departing at 3.31 p.m. This thirty-eight-minute stay was even faster than her first, in spite of transferring a greater number of passengers and mail. The only sad note of the brief visit was that all of the flags in the harbour were at half-mast in memory of William Watson, the Cunard Chairman who had been so instrumental in *Lusitania*'s construction and early years of service, who had died that very morning.[41]

While adding Fishguard to *Lusitania*'s eastbound itinerary, Queenstown was soon dropped as a port of call for *Lusitania* and *Mauretania* while eastbound. The following summer, Cunard announced that they had decided to cut out Queenstown as an eastbound stop for all of their express liners, not just *Lusitania* and *Mauretania*. The Irish simply wouldn't have that, and brought pressure to bear upon Cunard in the hopes that they would restore the stop. Cunard met them partway, restoring the eastbound stop for

41 *Evening Express and Evening Mail*, Tuesday, 5 October 1909.

Above: Lusitania's prow pointed towards the city of Queenstown, Ireland, during one of her visits to that port. (HFX Studios)

Left: The tender *Ireland*, so famous for its participation in *Titanic*'s first stop at Queenstown, tied up alongside *Lusitania* in Queenstown Harbour, apparently transferring mails. (Authors' Collection)

Lusitania made the first 'midnight sailing' from New York at the end of February 1912. Although they only lasted for about two years, these proved extraordinarily popular, a genuine social event for society in the era. (HFX Studios)

all of their express liners except for the sister speedsters. The Irish Parliament still was not happy. They wanted *Lusitania* and *Mauretania*, and continued to work on getting them back eastbound; meanwhile, Cunard continued to have the two liners call there while westbound.

Even with the new Fishguard stop eastbound, there were still times that *Lusitania*'s London-bound passengers did not arrive there until quite late at night. There was also the matter of the Irish screaming for the two sisters to resume their eastbound stops at Queenstown. Cunard wrestled with this conundrum for some time before making a remarkable decision: instead of having *Lusitania* sail at 9 or 10 in the morning from New York, why not sail her out eight or nine hours earlier, at 1 a.m.? Surely this would allow enough time for the sisters to stop at Queenstown again, and still get passengers to London before it was too late. At the eleventh hour, however, the decision to restore the eastbound stop at Queenstown was dropped; although the late-night sailing time was retained, passengers bound for Ireland would land at Fishguard and take the ferry back to Rosslare.

In December 1911 it was announced that both sisters would start this schedule at the end of February 1912, with *Lusitania* making the first of these so-called 'midnight sailings' from New York. Although Cunard hoped that the sisters would 'always sail at or near' that hour, the actual departure was left at the discretion of their Captains. If 'in their opinion the conditions should ever appear to be adverse', they had the power to delay and wait for improvement.[42] However, the Ambrose Channel was by then fully marked at night; all sorts of marine traffic was coming and going through the cut in darkness, so delays seemed likely to be the exception rather than the rule. Cunard was confident that 'midnight sailings' would prove popular, and they did. A buffet was served from 9 p.m. to midnight, just prior to departure. Many passengers, especially those booked in First Class, spent their last few hours ashore at social events and parties. Despite the popularity of these nocturnal departures from New York, in practice it was found that passengers bound for London still did not always arrive there before nightfall.

42 *The New York Times*, 28 December 1911.

During the summer of 1913, Cunard made the decision to cut all stops at Queenstown for *Lusitania* and her sister; Irish politicians renewed their protests with characteristic vigour, and delayed the actual implementation of this adjustment, although in the end their fuss was for naught. *Lusitania*'s last stop at the Irish port came at the beginning of Crossing No. 167, westbound, on 4 January 1914. Although she was scheduled to stop there on the outbound leg of her next round-trip voyage, on 1 February, rough weather prevented it. On 12 February, Cunard publicly announced that all stops at Queenstown by the speedsters were a thing of the past.

Even with midnight sailings from New York, no call at Queenstown and the stop at Fishguard with its express train service, Cunard took further steps to ensure that eastbound passengers would arrive in London at a reasonable hour. In March 1914, they pushed the 1 a.m. Wednesday departure time back to 6 p.m. on Tuesday evenings. This sailing time was only maintained through August of that year, and subsequently reverted to 1 a.m. before shifting back and forth between the two a couple of times. Finally, as a last change in her New York departure schedule, Cunard began sailing her out on Saturday mornings at 10 a.m. She remained on this schedule for the remainder of her career. Despite all these alterations to her schedule in departing New York eastbound, her westbound departure from Liverpool on every third Saturday never changed.

Occasionally *Lusitania*'s routine was interrupted by the unexpected. One good example of this came when *Lusitania* arrived in Liverpool on 15 August 1911. The city had been plunged into chaos for days; widespread labour unrest had led to strikes, rioting and mob scenes. *Lusitania* could not be docked because the tugboat men were among those striking; neither could she victual or refuel immediately. She could not be readied in time for her next scheduled departure, and was not ready to sail again until the evening of Sunday, 27 August; to help catch up on this catastrophic twelve-day delay the ship was run flat out, making three full crossings by 17 September. Turnarounds in New York and Liverpool between these crossings were compressed to shockingly brief affairs, and the turbines were not given a proper break. When the ship tied up in New York at the end of that third crossing, she held a new record for making three back-to-back crossings in three weeks. While there, the turbines would finally be given a proper chance to cool off, and inspected for damage due to the unusual period of sustained use they had endured.

Top: Passengers pose on the Sun Deck on 6 November 1909. The view is taken looking aft, with the Entrance skylight visible at left, and the Marconi Shack just behind that. From left to right the passengers are named Jack Foise of New York, Mrs Langdon, Alfred Langdon, and Mr Royse of Manchester. (Stuart Williamson Collection)

MIddle: The Second Class Boat Deck, looking forward. The portable weather screens for the Verandah Café are in place, while passengers engage in a game of shuffleboard. The weather seems to be chilly, as some of the men are wearing heavy overcoats and hats. (Günter Bäbler Collection)

Bottom: This spectacular photo was likely taken in summer, as the canvas screens are in place on the port Bridge wing. It was taken from the Stern Docking Bridge, and again shows the cranes stored pointed inboard. The portable weather screens for the Verandah Café are in place, while a later addition to the ship in this area, the 'bus stop'-style benches on the forward Second Class Boat Deck, are in place. The structures around the benches provided some shelter from the wind. (Günter Bäbler Collection)

Right: Seen from the starboard Bridge wing, passengers are enjoying the Sun and Boat Decks on a brilliant day. (J&C McCutcheon Collection)

Below: First Class passengers promenade along the Boat Deck. This view was taken on the port side and looks aft. In the foreground, a passenger enjoys the Atlantic vista from a wicker chair. (Authors' Collection)

Below right: This rare photograph shows a crewman posing with two Second Class passengers on the starboard side of the Boat Deck. Behind them, the two cranes are again stored inboard in what seems to be further evidence of an ongoing practice at sea. (Steven B. Anderson Collection)

On Wednesday, 6 December that same year, there came another schedule disruption: a gale in Liverpool caused *Mauretania* to break loose from her moorings and ground in the Mersey. She was refloated late the next morning, but Cunard decided that she should have a thorough hull inspection before her next sailing. Because of this, *Lusitania* was given the honour of the pre-Christmas mail crossing instead. She departed Liverpool on Saturday, 8 December and arrived at her New York pier at 7 a.m. Friday, 15 December; after a high-speed turnaround there, she made the eastbound crossing, arriving back at Fishguard, Wales, just twelve days and seventeen hours after leaving Liverpool.

Technical problems also occasionally disrupted *Lusitania*'s regular service schedule. One of these transpired at about 8.30 on a Monday morning in late March 1909. Just a day out from Queenstown, the ship began to shudder violently for about four minutes. Concerned passengers left their breakfasts or came up from their staterooms in disarray, hoping to find the cause. Down in the Engine Room, the reason was immediately apparent: one of the propellers had thrown a blade. To prevent damage to the machinery, all engines were immediately stopped. Then, to discover which prop was the offending one, the engines were restarted; the inboard starboard prop was soon identified as the one that had suffered damage. The day after she arrived in New York, divers were sent down to inspect the nature of the trouble. It was worse than originally believed. One of the three blades had broken off close to the head, and another had sheared at mid-blade. A new pair of propellers would need to be installed, but it could wait until the ship returned to England. She made her next crossing eastbound on only three propellers; the new set that replaced the older pair was a marked improvement, being more like *Mauretania*'s high-performance screws, and improved *Lusitania*'s performance.

More tiresome than this rather commonplace mishap, however, was a protracted series of difficulties with *Lusitania*'s turbines. The troubles began in June 1912, on Crossing No. 141 West. One of the passengers was Mrs Joseph Loring, a widow of one of the victims of the *Titanic* disaster just two months before. When the liner was ready to depart Queenstown on Sunday, 9 June all seemed well; the order was rung down to the Engine Room for 'Half Ahead' on the low-pressure turbines that drove the inboard props. The engineering staff put steam pressure on the pair, but the port turbine wouldn't budge, not even with 20,000hp applied. There was no easy way to find the cause

Three passengers sitting on the Boat Deck behind the Second Class Lounge. The Stern Docking Bridge rails are reflected in the windows. (Authors' Collection)

Antarctic explorer Ernest Shackleton and his wife Emily arrive in New York on 25 March 1910, at the conclusion of Crossing No. 75 West. The couple stand on the port Boat Deck outside the Verandah Cafe. Shackleton had been knighted by King Edward for his efforts to reach the South Pole the previous January. (Library of Congress, Prints & Photographs Division, Authors' Collection)

This rare photo was taken on the port-side Boat Deck. The woman's name, Marie, was written on the margin of the photo. On the left are the windows of the Reading and Writing Room, while on the right, the door leading into the First Class Entrance is open. (Stuart Williamson Collection)

Top: Crewmen offloading the Christmas mails in New York, December 1910. The post was always a vital factor in the ship's career, as was the mail contract that allowed her to carry the designation 'RMS', or Royal Mail Steamer. On Crossing No. 31 West, 3–9 October 1908, *Lusitania* even had the distinction of carrying the first penny postage from Liverpool to New York. (Library of Congress, Prints & Photographs Division, Authors' Collection)

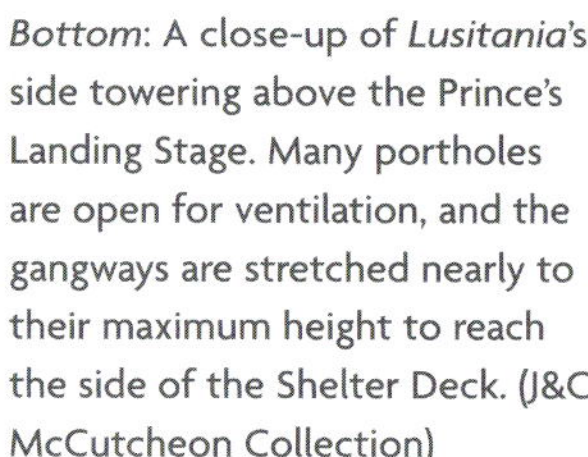

Bottom: A close-up of *Lusitania*'s side towering above the Prince's Landing Stage. Many portholes are open for ventilation, and the gangways are stretched nearly to their maximum height to reach the side of the Shelter Deck. (J&C McCutcheon Collection)

of the dilemma; answers would not be forthcoming without lifting the cover off the turbine. It was decided to hold off on repairs until the ship returned to Liverpool. However, the prop was frozen in place and creating drag on the ship as she moved through the water. By Sunday afternoon, the Chief Engineer decided to de-couple the shaft at the neckpiece so that the prop could spin freely in the water like a windmill, reducing the drag. After the engineers removed all but two of the connecting bolts, the vessel was stopped. Once the last two bolts had been removed, the ship resumed course, and it was reported that the prop spun freely at about 100rpm all the way to New York. When *Lusitania* arrived in New York, five days, twenty hours and fifty minutes after leaving Queenstown, her average speed had only been 21.48 knots. She returned to Liverpool at a similarly slow pace, and her next crossing was cancelled while her turbine cover was lifted to discover the source of the problem.

Lusitania in drydock around 1911. The weather is cold, as long overcoats are being worn by the gentlemen walking in front of the ship. (Museum of New Zealand, Authors' Collection)

Following this incident, *Lusitania* made a round-trip voyage in July all right, but on the following westbound passage, spanning late July through early August, there came another episode. While en route from Liverpool to Queenstown in heavy weather, at about 9 p.m. Saturday, 27 July, it was discovered that the bearings of the port high-pressure turbine had overheated until the brass cap had begun to melt. The bearing – which weighed 1½tons – would have to be removed and replaced with the spare, which was stowed all the way forward by the Firemen's Quarters. Extra men from every department were sent into the titanic effort as, by a system of pulleys, the replacement bearing was taken out of storage and slowly moved over to the port side. Then it was moved aft until it could be hoisted into position to replace the faulty bearing. The task took a full twelve hours; by noon on Monday, the ship had made only 469 miles. Although the new bearing worked all right, her average speed for the crossing to New York was a mere 21.19 knots.

A second photo from the same stay in drydock shows the ship's quartet of four-bladed propellers, which helped improve her speed and vibration issues. (Museum of New Zealand, Authors' Collection)

With a second turbine experiencing catastrophic failure, it was clear that something was amok with *Lusitania*'s engines. Beginning on 15 October 1912, she was removed from service for nearly two full months. John Brown was tapped to open up all of the turbines with the exception of the starboard high-pressure one, and to carry out inspections and repairs on the equipment. Cunard invested £7,330 in this work, and the ship re-entered service, departing Liverpool on 13 December on her seventy-seventh round-trip voyage.

During this westbound crossing, on 19 December, she averaged 27 knots – including a one-hour burst of nearly 28 knots, turning out 207 revolutions on the engines – for a full day of steaming. While she did not manage to recapture the Blue Riband, it was an impressive feat, and the turbine work may have seemed eminently successful. She returned to Liverpool, completing the round-trip voyage on 31 December, and everything seemed all right to the public. Then came a surprise: almost as soon as the liner had tied up, Cunard announced that her next crossing was cancelled. When enquiries were made regarding the cause, Cunard responded simply that there had been a 'slight derangement in the machinery'.[43] That was all the public was told.

Behind the scenes, the situation was quite serious. Fresh from two months of significant turbine maintenance and repairs, another serious turbine problem had occurred. It had happened as *Lusitania* was approaching Fishguard Harbour on the afternoon of Monday, 30 December. A steamer coming out of Fishguard had come too close for comfort, and to avoid a collision, the helm was put hard to port. This attempted turn to starboard failed when the telemotor jammed, and *Lusitania* remained stubbornly on course. By then the situation was quickly becoming critical, and the order was given for 'Full Astern' – without stopping the turbines before changing over to astern thrust. The engineers complied, and a collision was avoided. Once the situation had calmed, it was found that the steering gear had failed because a bit of marline had dropped into the telemotor, jamming the system. If everything else seemed all right, that illusion was quickly shattered. When the time came for the ship to resume her trip towards Liverpool, the port low-pressure turbine – the blades of which had just been replaced – would not move. *Lusitania* limped back to Liverpool on only three props; tugs met her at the mouth of the Mersey to help her manoeuvre the tight confines of Liverpool waters. A preliminary investigation revealed that the blades of the port low-pressure unit were twisted out of pitch. Additionally, the blades on both of the forward-thrust turbines on the starboard side had also been damaged.

Obviously, repairs were going to take some time – but just how extensive they would have to be was not yet clear.

43 *The New York Times*, 1 January 1913.

A view in late December 1912 between Crossings Nos. 153 West and 154 East. It was *Lusitania*'s first post-turbine refit round-trip voyage; during the lay-up, some of her original ventilators had been replaced with cowl-style ones more like *Mauretania*'s. At the end of the upcoming return trip to Liverpool the engines would be so severely damaged that *Lusitania* would be out of service for eight months. This was also the first crossing she had made without Purser Lancaster, as he had just retired. (Library of Congress, Prints & Photographs Division, Authors' Collection)

The ship's forthcoming voyages were cancelled, and she was replaced by shifting about other vessels in the fleet. The crippled *Lusitania* was soon placed in the Canada Dock so that the full extent of the damage could be ascertained. It was hoped that the blades on the port low-pressure turbine could once again be replaced, and that the damaged blades on the other turbines could simply be straightened out without having to replace them. Just investigating what needed to be done was an enormous task, involving raising the turbines' casings to ascertain the damage, encompassing two full weeks. What they found was a real mess. It would be impossible to simply patch up the turbines; they would instead be forced to completely re-blade and service all four of the ahead turbines. The contract for work was awarded to Cammell Laird & Co. and Parsons & Co.; for some time *Lusitania* could be seen in the Canada Dock in Liverpool. She did not return to service until late August 1913, eight full months after the Fishguard incident.

These repairs cost Cunard a total of £7,639. Some £2,432 of that figure was spent opening up the port and starboard astern turbines, inspecting them and putting them back together with no repairs made. Other work carried out on the ship during this extensive eight-month overhaul brought the total expenditure to over £50,000. This was an astounding sum, especially when one considers the revenue lost in the liner's absence from service – which some estimates put at £500,000. Cunard was also not particularly pleased that John Brown's 1912 repairs, which had cost over £7,000 in their own right, had not held up for more than a single round-trip voyage. Negotiations between the two firms commenced on the charge for the second set of repairs, and were concluded amiably by mid-October 1913.[44]

44 *The New York Times*, 23 August 1913; Cunard Archives, University of Liverpool.

A stunning view of *Lusitania* in the new Gladstone Graving Dock in Liverpool, which had been opened in July 1913 by King George V. Amidships, workmen can be seen installing the new lifeboat davits amidships; another pair can just be discerned at the edge of the Second Class Boat Deck. It was apparently around this time that the mounts for *Lusitania*'s 6in guns were installed on the Forecastle, although the guns themselves were never installed; also, the aft set of cranes – removed just prior to the maiden voyage in 1907 – were put back in place and remained there through the rest of the ship's career. (J&C McCutcheon Collection)

In a fascinating twist, an opportunity was taken during the 1913 turbine repairs to reverse the direction in which the outer, forward, propellers turned; whereas up to that point in the ship's career, this set of props had rotated inward, while the turbines were re-bladed their direction of rotation was changed so that these props would now rotate in the outer direction. This was said to be based on experience with *Mauretania*'s performance, and hopes were high that it would result in her recapturing the Blue Riband. Although these hopes were not realised, this little-known alteration meant that for the first time, all four props would spin in an outer direction, with the two props on either side spinning in the same direction for the first time in her career.[45]

These large-scale disruptions to *Lusitania*'s schedule were rare, however; more common were the less troublesome but always noteworthy occurrences that fell outside the purview of 'ordinary'. For example, the January 1908 departure from New York became a scene of utter chaos. The first problem was that eager ticket agents, looking for commissions selling cabins, had overbooked the Cunarder's accommodation. Because of this, a great crowd of would-be Third Class passengers had to be turned away at the gangplank. Shortly thereafter, word was passed down that there were still some Second Class cabins available for an upgrade of only $13. The crowd rushed to the ticket office on the pier en masse, clamouring to take the upgrade. While beleaguered employees sorted through the mess, *Lusitania*'s whistle sounded; a horse pulling a cab on the pier started at the unearthly noise, and charged right towards the horde by the ticket office. Traffic policemen warned the crowd, which scattered quickly. One Italian gentleman, however, could not get out of the way in time and was knocked down; fortunately, he suffered nothing more than a bad bruising. A traffic officer eventually stopped the frightened beast, and its driver was arrested for not having a proper cab licence.

Sometimes commotion came in a completely different manner. In July 1908, a well-dressed and obviously well-to-do First Class passenger was unceremoniously escorted from

45 *Bath Chronicle and Weekly Gazette*, 16 August 1913. The article's statements have been confirmed through a careful photographic examination of the ship's propellers – the three recovered from the wreck in 1982 and the fourth, which remains in position on the wreck, and which just happens to be one of the two props that rotated outward after the 1913 turbine repairs. The blades on the prop prove that it had been changed from its original direction of spin, and that the change made in 1913 had been retained until the end of the ship's career.

Lusitania before she departed New York. It seems he had caused a sensation soon after boarding; once comfortably ensconced in his stateroom, he made an aberrant attempt to call up the Tsar of Russia, the King of England and the Pope in Rome to inform them all that he was on his way to pay them a visit. There were some lines of social behaviour that one just didn't cross; Chief Surgeon Dr Pointon was consulted, and suggested that the man's health was not robust enough to stand the ocean voyage – very likely an exaggeration, but a convenient loophole nonetheless. As the man was being escorted from the ship, he was very agitated, but after his ticket price was refunded he left quietly.

There were the eccentric, and then there were the outright lawless. Customs inspectors in New York were renowned for their ability to sniff out passengers who felt adventurous enough to lie on their customs declaration forms, or who attempted to smuggle items into the country on their person. Despite the inspectors' vigilance, some people still tried to bend the rules, with varying degrees of success. In June 1910, four women travelling together in First Class failed miserably at this charade. First they stated that they had nothing whatsoever to declare. This immediately roused suspicions, and an examination of their nine steamer trunks and hand baggage was set to begin. At the last moment, they changed their story to a declaration of $160. This was just too much to believe, and they were sent back aboard the ship for a thorough examination of their person. Female inspectors were called in to carry out the examination, and emerged with a veritable treasure trove of about $3,500; they found nine items of jewellery in a bag strung on a girdle about the waist of one woman, and other items were soon found on the three other women. All four of these female travellers were wealthy – one of them was even married to former Chicago Mayor DeWitt Cregier (term 1889–91). They were all taken to the Federal Building for further questioning. Despite the damning evidence found on their person, all four women strongly denied that they had been attempting to defraud the Government.

Other passengers were suspected of having perpetrated far more serious crimes. One particularly profound incident swept *Lusitania* up into the middle of an international manhunt. It all began on Saturday, 26 December 1908 in Cunard's Liverpool offices on Water Street. With *Lusitania* due to sail for New York that day, there were many prospective passengers who needed attention from the office personnel. At about 12.30 p.m., a slim-built gentleman wearing a vicuna overcoat and top hat for protection from the cold entered the office. Soon he was assisted by John Forsythe, Manager of the Second Class department. He requested accommodation for him and his wife aboard *Lusitania*. Forsythe pulled out a diagram of the liner's passenger accommodation, and offered the man inside Second Class cabin E76 for £12.

The gentleman did not seem particularly happy with the offer, explaining: 'No, I do not like that, it is inside; it was offered by your agents in Glasgow.' When he made the remark about the Glasgow agency, he seemed to immediately regret having mentioned it, and immediately added, 'by your other agents'.

Forsythe thought this a little odd, but went back to the drawing board. Soon he had found an alternative, and offered cabin C1, which was just forward of the well overlooking the Second Class Dining Saloon, and also had a view looking out on the aft portion of the First Class deck. The price for this cabin was £28. Again, the gentleman was not particularly happy, this time with the price.

The Cunard agent stuck to his guns, however, and soon enough the man acquiesced and accepted cabin C1 for the £28. He paid with six £5 Scottish notes, and was given £2 in change. Forsythe asked his name.

'Otto Sando. S-a-n-d-o. It is not Sandow, the strong man,' he said with a jocular smile.

Forsythe gave him an application form for tickets, No. 37/5, which the man filled out carefully. He gave his wife's name as Anna, his age as 38 years, and his profession as a dentist. He put down that he was from Germany, but was also a US citizen and gave a Chicago address.

The Cunard agent noted that Sando seemed a trifle nervous; at one time when the office door opened, Sando looked expectantly in that direction. It wasn't the first nervous passenger that Forsythe had seen, however, and he dismissed it from his mind. After exchanging another small amount of the man's money, Forsythe watched him leave the Cunard office to return to his hotel.

The Sandos boarded *Lusitania* and the ship soon cast off. The weather during the westward passage, the outward leg of Voyage No. 20, was decidedly foul, however – 'very rough', as Second Class Bedroom Steward Thomas Atherton recalled. Second Class Stewardess Beryl Bedford recalled that she was very busy with the passengers in her care, as many of them, including 'a great many ladies', were seasick. Anna Sando stayed in her cabin for most of the trip, although her husband made it down to the Dining Saloon for his meals. He told Steward Atherton that he had crossed the Atlantic a half-dozen times and 'had not previously been sick'.

Apparently, the Sandos were not pleased with the service that they received during the trip. Rather early in the voyage, Stewardess Bedford heard from the Chief Steward that they had complained of a 'want of attention' to Mrs Sando during her illness. The stewardess must have chafed at the complaint, no doubt being run ragged by the many ill passengers she was dividing her time between. She did not, however, confront the Sandos about their complaint when she returned to assist them, nor did they complain about the matter to her directly.

When the voyage ended on the morning of 2 January 1909, it must have seemed like just another run-of-the-mill winter crossing, and Stewardess Bedford was doubtless pleased that the Sandos would soon be disembarking. However, as *Lusitania* was picking her way up towards her pier, a revenue cutter approached and several lawmen and detectives boarded. Something was amiss. It turned out that they were on a manhunt, looking for a 'slim man, with a dark complexion, about 5 feet 7 inches in height, 35 years old, and with a crooked nose'. When they came across Otto and Anna Sando, they found that he met the description of their wanted man. He was suspected in connection with the 21 December murder of Miss Marion Gilchrist, an elderly lady who had lived in Glasgow.

The man quickly admitted to the authorities that 'Sando' was a pseudonym, and that his real name was Oscar Slater. As the ship travelled up the Bay and tied up at her pier, he and his wife were questioned. Slater admitted that they were travelling from Glasgow; when asked if he knew the deceased woman, he said he had never heard of her. His wife agreed that she had not known of the woman either. Slater was next formally arrested, and when his person was searched, a pawn ticket from a Glasgow broker was found. It was for an advance of £60 on a three-rowed crescent diamond brooch, pawned on the day of the murder; the brooch was known to have been stolen from Miss Gilchrist by the murderer. Slater told the authorities that the brooch he had pawned was a different one than Gilchrist's, and that it had been in his possession before the murder had even taken place.

Quickly realising the gravity of his situation, and seemingly 'becoming confused, [Slater] added that he knew the woman was murdered because he had read about it in the papers before he left Liverpool'. The authorities felt that the pawn ticket was 'significant' evidence, and took Slater to the Tombs detention complex to set a hearing. His wife was taken to Ellis Island for further examination. Before she disembarked *Lusitania*, she gave this statement to the press, speaking with a 'slight French accent':

> I was married to Mr. Slater in Glasgow on July 12, 1901. My maiden name was Miss Andrea Anton. I was a daughter of Malone James Anton, and lived in Paris. I came to America with my husband soon after the marriage. Two years ago we left here on the *Kaiser Wilhelm II*. We spent most of the time in France until three months ago, when we went to Glasgow, where my husband then established himself as a dentist. I myself have been to this country several times in my life.

Back in Slater's cabin, C1, Bedroom Steward Atherton began cleaning up in preparation for the next batch of passengers who would occupy it. Therein, he found a copy of the previous Sunday's *News of the World* newspaper, which was frequently brought aboard the vessel at Queenstown. Atherton picked up the newspaper, curious to see if he could find some mention of the Gilchrist murder in its pages, no doubt surprised at how he had suddenly and unwittingly been drawn into the case. He found that the article about the murder had been torn out of the paper; thinking little of it, however, he threw the newspaper away and continued to clean the cabin.

Oscar Slater's story continued to unfold over time, gathering attention from the press and public alike. Although there was not enough evidence to warrant extradition back to Scotland for a trial, Slater agreed to it, apparently believing that he could easily clear his name of the charges. Stewardess Bedford and Bedroom Steward Thomas Atherton were called to give witness in the subsequent trial, a trial that many believed was biased. Slater was found guilty of murder in May 1909 and he was condemned to death. Because of public outcry, however, his sentence was commuted to life in prison.

A year later, a Scottish lawyer and amateur criminologist, William Roughead, published the book *Trial of Oscar Slater*, which showed clearly the flaws in the prosecution's case, including the irrelevance of the 'significant' pawn ticket from Glasgow. In 1912, Sir Arthur Conan Doyle himself – famed creator of the fictional detective Sherlock Holmes – became involved in clearing the man's name. He published *The Case of Oscar Slater*, in which he pleaded for a full pardon. *The Truth About Oscar Slater* was published by William Park, and by July 1928 the conviction was overturned. Slater was released and received £6,000 in compensation for the years of mistaken imprisonment.

Indeed, Slater's trial and imprisonment was a gross miscarriage of justice. Although he was no saint, as the saying goes, he certainly was not guilty of murdering Miss Gilchrist. Slater was a German-born Jew who had lived for some time in the United States. When he moved to Glasgow and set up as a dentist, it soon became clear to people in the community that he was neither a dentist nor a man of high reputation. He had already been married in 1901 or 1902 to a woman named May Curtis, but later separated from his wife, apparently because of her drinking problem. Apparently, she subsequently tried to keep in touch with him, 'always bothering him' and looking for money. Slater lived his life as a club manager, and was also apparently involved in trading precious stones, associating with criminals in the process.

Around 1905 Slater had met a woman named Andrée Junio Antoine at the Empire Theatre in London. At the time, she was around 18 years of age, but was already a prostitute, with the professional name of Madame Junio. Soon a romance bloomed between them, and they found a new life in America, travelling there under assumed names, where he ran a club in New York City. It was said that they formed a 'left-handed alliance', but whether this was a formal marriage or a common-law one is uncertain. Eventually, they returned to Europe, settling in Glasgow only a few months before the murder. Although he was not entirely of good reputation, Slater had never been on the wrong side of the authorities before being charged with the slaying. Numerous witnesses vouched for his innocence, including Miss Antoine, who provided an alibi for his whereabouts at the time of the murder. The couple's departure from Glasgow had nothing to do with fleeing the murder scene, but was instead planned for some weeks, and had more to do with finding someone who agreed to take their flat than it did with guilt. They were travelling under assumed names, just as they had before, in order to avoid detection by Slater's wife, who was apparently still looking for him. It was later concluded by impartial – and informal – investigators that Miss Gilchrist was murdered by her own nephew.

Slater's case was a sad miscarriage of justice. However, the connection between this notorious story and *Lusitania* was remarkable and was perhaps the most incredible tie to a criminal investigation in the whole of her career. The connection between the Slater case and the legendary liner, however, was little-remembered in subsequent years.[46]

Lawbreaking wasn't confined only to *Lusitania*'s passengers. For several months in 1910, there was a suspicious rash of robberies aboard. On crossing after crossing there were complaints from passengers of being robbed; on one trip alone some $4,000 came up missing. There was no chance that the robberies could have been by a single passenger, because it was spread out over such a protracted period of time. Finally, that November, the perpetrator was caught: it was Boatswain's Mate Joseph Weavill, who had been with the ship since her maiden voyage. Weavill had moved through the ranks from Seaman to Quartermaster, and then by his own request to Boatswain's Mate. This particular job put him in a position to commit the robberies. He was arrested in Liverpool and sentenced to six months in jail.

Captain Turner was also apprehended once, during the late summer of 1909, although for a far less nefarious reason. Turner's beloved niece, Miss Mercedes Desmore, was then living in New York, and one of Turner's favourite pastimes was to be a good uncle and spoil her at every opportunity. She boarded the liner to meet her uncle one Friday night, and when they disembarked at 6.30 p.m., she was carrying a gift from Turner: a half-pound package of chocolate. As they passed the customs gate, she happened to be holding it in her hand and eating a piece of it. One of the customs inspectors stopped them dead in their tracks and insisted that they

46 The information on this story was taken from a variety of sources, including *Trial of Oscar Slater* (1929) by W. Teignmouth. Within that volume are transcriptions of original sworn testimony from the witnesses called for the case, including that of Cunard Agent John Forsythe, Stewardess Beryl Bedford, Bedroom Steward Thomas Atherton and of Slater's partner, Andrée (or Andrea, or Anna, depending on the account) Junio Antoine. Information on the arrest aboard *Lusitania*, including Miss Antoine's statement to the press, was gleaned from *The New York Times* of 3 January 1909.

Certain questions regarding Slater's history, and his relationship with Miss Antoine, remain unanswered. Although many sources cite the date of his first marriage in 1902, some place it in 1901; Miss Antoine said in her statement to the press that she married Slater on 12 July 1901. Since Miss Antoine did not meet Slater until much later, was this perhaps the actual date of his marriage to his first wife? There are other questions, as well, worthy of a full investigation by modern researchers. The basic facts of his case, however, seem quite clear, as is Slater's innocence insofar as the murder of Miss Gilchrist.

could not leave until a duty had been paid on the offending chocolates, even taking them right from the girl's hand. Turner was incensed. During his entire career with Cunard he had never been stopped at the customs gate; to be treated like a criminal in front of his niece – especially over such a trivial thing – was unbearable. He quickly stormed off and reboarded *Lusitania*, but left a profound impression on the zealous customs official. Later on the customs man decided to return the chocolates to Captain Turner with an apology.

Somewhat less humorously, in early September 1910, there was quite a scare on board. The ship arrived in New York late on a Thursday night with a nearly full complement of passengers; demand for this particular crossing had been very high. After making her customary call at Quarantine and being released, she set her nose towards her West Side Pier. Over a thousand people were gathered there to meet disembarking friends and relatives. Only about fifteen minutes after receiving clearance, word came for *Lusitania* to return to Quarantine immediately. A Romanian passenger in Third Class had a disease that the Deputy Health Officer could not quickly diagnose. Over an hour passed before a simple communiqué reached the city:

> Am anchoring off Stapleton for the night. Have not been passed by the doctor. CHARLES.

There were 915 First and Second Class passengers and 1,125 in Third Class. As word passed around the ship, concern grew, as there was a cholera scare in Europe at the time; no one wanted to be stuck on board with an ailing passenger if there was any possibility of becoming sick. Overnight, bacteriological tests were carried out, and by morning they came back negative for anything infectious. The ship was duly cleared at 9.20 a.m., and docked just before 11. Although the scare was real, the threat had fortunately not materialised into anything dangerous.

Corporate espionage was also a constant reality, particularly during 1907 and 1908. When *Lusitania* and *Mauretania* entered service, they were far and away the most technologically advanced liners of the time, and the German steamship lines were desperate to find out more about their workings. Armed with such information, they could improve their next generation of liners and also learn whether such liners were financially feasible without government subsidies. In early September 1908, there was a group of steamship men on the floor of the New York Maritime Exchange, and their conversation turned to whether the new Cunarders were really turning a profit. The general feeling seemed to be that if the British Government had not been providing an annual subsidy, they would have been losing ventures.

At this, a representative of a German steamship company piped up. 'I will show you this is not so,' he said. To back his claim, he produced a detailed set of figures that showed just how closely his company had been studying the new British ships during the preceding year. He was able to prove that not only were the Cunarders making good money, but that the entire Cunard fleet had picked up traffic because of the company's boasting rights at having the world's largest and fastest liners in their proverbial stable.

The Germans had certainly done their homework. During 1908, a Second Class passenger was seen travelling on *Lusitania* quite frequently. He was in fact a German marine engineer employed by one of the German steamship companies. For some time this engineer had been known to possess a developed ear for listening to the operation of ships' engines. It was said that he could tell by the engines' strokes just how much coal was being consumed, and that every sound had a definite meaning for him. After months of travelling aboard, with his ear glued to the bulkheads, he reported to his superiors in their New York office that she was more efficient than commonly supposed. The Germans did not rely on ears alone. It was said to be rather common during late 1907 and through 1908 for stokers, oilers and firemen to be thrown off *Lusitania* and *Mauretania* after evidence had surfaced that they were German spies trying to find out more details of the liners' technology.

Trying to steal a ship's secrets would hardly have caused a threat to a liner's safety; however, in November 1911 such a threat did transpire. Shortly after departing Queenstown for New York, electrical light wires in a Second Class cabin short-circuited and started a fire. The crew extinguished it in short order, before it got out of hand. Initially the press reported that passengers were driven to the decks by 'thick, pungent smoke', and that the lights in the Second Class regions of the ship had quickly failed. In reality, the affair was far less dramatic; First Class passengers were not even aware of the problem until the ship had docked in New York.[47] *Lusitania* suffered from several other fires throughout the later years of her career, as well, but happily each of these was extinguished before getting out of control.

47 *The New York Times*, 21, 25 November 1911.

Less dangerous to the ship's safety, but nevertheless an irritation, were the professional gamblers looking to prey on gullible First Class passengers. One period guidebook advised: 'Gentlemen should be very cautious about playing cards, or other games, with strangers, as professional gamblers are constantly crossing the Atlantic, looking out for the unwary. There is nothing unusual in the captain posting a notice in the smoking room warning passengers against gamblers.'[48] Such notices were displayed prominently aboard *Lusitania*, and at times special warnings were also posted. In January 1911, when one well-known cardsharp was spotted boarding with a companion, the two men

48 *The Scientific American Handbook of Travel*, p.115.

Right: Lusitania tied up at Pier 56 during the Hudson–Fulton Celebration in late 1909. She is dressed out for the event, and grandstands were erected outside Pier 56 so spectators could watch the river traffic coming and going. The *Half Moon* can be seen near *Lusitania*'s stern. (Library of Congress, Prints & Photographs Division, Authors' Collection)

Below: Lusitania departs New York on 29 September 1909, during the Hudson–Fulton Celebration. (Courtesy Digital Culture of Metropolitan New York)

Above: Wilbur Wright prepares to take off from Governor's Island to fly over the departing *Lusitania*. (Authors' Collection)

Below: The 4ft-tall single-chime whistle on *Lusitania*'s forward funnel being blown while a crewman works on the port-side siren. The two single-chime whistles on the forward funnels made quite an impression on the pioneering aviator as he flew overhead; he said he could feel the whistles' pounding vibration resonate through his body. (Authors' Collection)

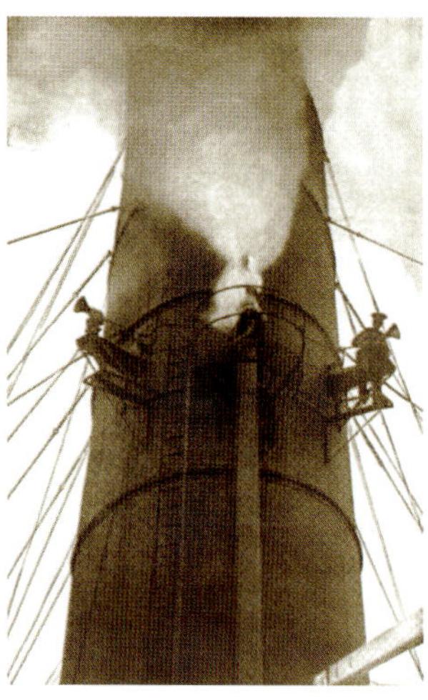

were confronted by the ship's officers. They reassured the officers that they were not aboard for work, but had instead undertaken the crossing for other reasons. Captain Charles was informed immediately; he ordered that the men's descriptions be printed and posted in the main companionways so that passengers would be forewarned and – it was hoped – refrain from playing games of chance with them.

Even these precautions did not always protect passengers from being duped – warnings could be posted, but if passengers decided to play, there was little that Cunard could do to protect them. That is exactly what happened late that May: a pair of professional gamblers took an estimated $14,000 – over $449,000 in 2023 – during westbound Crossing No. 111. At times, over $1,000 had been sitting in the pot; although it did not become widely known what was going on, some took an interest in the gaming and sat in to watch events unfold. One man played for an hour, losing $1,000 for his efforts before he quit in disgust at his 'ill-luck'. Luck had nothing to do with it. One passenger who was observing had lived in the western United States – an area as notorious now as it was then for gambling and cheating – for most of his life. He recalled that he had 'never in his experience … seen such quick action and such a passing of signals as had marked' this series of games. The marks were dealt very good hands, 'hands which, had the game been on the level, would have been worth every bit of the risk taken upon them'. These 'good hands' caused the players to wager in full confidence; at each turn of the cards, however, an even better hand prepared by the professional gamblers would top them. The losers paid up in full, and the card-sharps disembarked with their enormous take.[49]

While some incidents were troublesome, other events proved to be fascinating or even pleasant. Between 25 September and 11 October 1909, the State of New York commemorated the 300th anniversary of Henry Hudson's first exploration of the Hudson River in 1609, and the anniversary of Robert Fulton's first voyage on that river under steam in 1807. The festivities were aptly christened the 'Hudson-Fulton Celebration'. British and US warships anchored in the North River to participate in the event. On the day before the occasion began, *Lusitania* arrived in New York. Steaming towards her pier at about 4 p.m., she came up on the HMS *Inflexible*, which had not yet reached her anchorage. The pilot kept *Lusitania* quite close to the British flagship, and the warship's band serenaded *Lusitania*'s passengers with British and American patriotic airs as they steamed side by side for two hours. The warship's decks were manned in salute, and *Lusitania*'s passengers cheered wildly; finally, *Lusitania* pulled ahead and proceeded to her pier.

Five days later, on Wednesday, 29 September *Lusitania* was preparing to depart New York amidst the same celebration. That morning, famed aviation pioneer Wilbur Wright made three short but groundbreaking flights around New York City, taking off from Governor's Island. These flights became a real highlight of the overall event. During the second of these adventures, which took place just after 10 a.m. and lasted just five minutes, the audacious Wright passed directly over *Lusitania*, which had just departed her pier and was moving down the North River towards open water. Wright had received many salutes from waterborne craft that morning, but it was *Lusitania*'s pair of 4ft-tall brass single-chime whistles that really made an impression on him. He later recalled that even from high in the air, he could feel their pounding vibration resonate through his body. The surreal moment passed quickly, but it was certainly noteworthy in historic context: just fifty short years later, aircraft would sound the death knell for transatlantic passenger trade.

Lusitania also seemed to be gifted with the romantic touch. Because of her prestige and her proximity to New York and London society, it was quite common to find newlywed couples aboard, travelling on their honeymoon. It was

49 *The New York Times*, 27 May 1911.

A romantic walk on *Lusitania*'s moonlit decks fuelled many a shipboard romance during her career. (HFX Studios)

Lusitania sailing westbound into a stunning sunset. The great liner was always gifted with the romantic touch, and was a favourite of honeymooning couples. (HFX Studios)

In this still from a film shot aboard *Lusitania* during a westbound crossing *c*. 1911, a man stares out from the Second Class railing over the ocean. (Authors' Collection)

Crewmen perform maintenance work on the lifeboat during the same 1911-era westbound crossing. (Authors' Collection)

considered quite the thing to have an early morning wedding in one of New York's notable churches, to take breakfast at Delmonico's and then to embark aboard *Lusitania* for her eastbound crossing to commence the honeymoon. Every wedding day has its stresses, however; in October 1908, Mr Forbes Hennessy and Miss Margaret Sheehan had quite a shock when they found that *Lusitania*'s departure had been moved up from 11 a.m. to 9 a.m. The ceremony had been scheduled for 8 a.m. that morning at St Patrick's Cathedral, with breakfast at Delmonico's to follow. The three-month-long European honeymoon simply wouldn't wait, and it was decided to move the ceremony up to 6.45 a.m. The officiating monsignor was no doubt trying to stifle a yawn or two as he presided, and once the couple had been officially joined, they hastened directly aboard the ship, choosing to breakfast in *Lusitania*'s First Class Dining Saloon instead of Delmonico's. Here the cuisine and the surroundings were no doubt just as splendid as at the originally intended location.

Crossings aboard *Lusitania* sparked frequent 'shipboard romances', as often happened aboard other ships. Such fraternisation was almost always confined to members of a single travelling class, however, and not between those from opposite ends of society. 'First class passengers are not allowed to enter second or Third Class compartments, and vice versa, as complications might arise under the quarantine regulations. Visits to the steerage can only be made by special permission', prospective travellers were advised.[50] However, within the confines of a single class, there were many shipboard romances aboard *Lusitania* in those years – whether it had to do with the invigorating sea air, the romantic atmosphere aboard or even just a moonlit stroll on the Promenade Deck, where cupid marked 'the spot and the scene for his own'.[51] Although most of these romances faded with the crossing's end, others thrived for years to follow.

Sea voyages were always social affairs, even when romance was not involved; 'shipboard friendships' are legendary with good reason. One guide to ocean travel put it this way: 'The forming of pleasant acquaintances passes away many agreeable hours … With discernment and tact often a lifelong friendship is formed on an ocean voyage.' By observing social amenities and courtesies, it was virtually guaranteed that the voyage would be 'one of the most agreeable experiences of life'.[52]

It may have seemed like the good times would never end; yet during her career, the world began to change around her. Although *Lusitania* was the largest 'in service' liner for less than two months, Cunard touted her and *Mauretania* as the world's largest and fastest steamships in their advertising, dividing the honour between the pair. After June 1911, however, they could only claim the title of the 'world's fastest steamships', for in that month, White Star Line put into service the newest and largest steamship in the world: *Olympic*.

Olympic was far larger than either Cunard sister. She was 882ft 9in long – some 95ft longer than *Lusitania*. She was also somewhat wider, with a maximum breadth of 92ft 6in. *Olympic* sported a fuller figure than the Cunarders, for she had been designed primarily for comfort at sea, not speed.

50 *The Scientific American Handbook of Travel*, pp.98–9.
51 *Outward Bound*, John H. Gould, p.22.
52 *Ibid*.

A view on 21 June 1911, as the new White Star liner *Olympic* arrived at White Star's Pier 59, just north of Cunard's Pier 54. After *Olympic* crossed *Lusitania*'s stern, recently knighted Captain Sir James Charles backed *Lusitania* out of her pier just in time for photographers to capture the two ships together. (Library of Congress, Prints & Photographs Division, Authors' Collection)

At 45,324 tons, she was nearly half as large again in enclosed volume as the speed queens, and her fuller beam supplied a higher rolling stability that was very forgiving for passengers. Being kept out of the race for the Blue Riband at a service speed just over 21 knots, *Olympic* also lacked the noticeable vibration of the Cunarders. She was powered by two four-cylinder, triple-expansion reciprocating engines driving outboard screws, with waste steam vented into a single low-pressure turbine driving a centre prop. *Olympic* was a true 'floating palace', with plenty of space for the niceties and comforts that the Cunarders simply could not afford within their slim hulls.

Lusitania and *Mauretania* were hopelessly outsized, but they had the advantage of a good reputation and a loyal following; their reputation for safety was highlighted when *Olympic*'s sister ship *Titanic* made her maiden voyage on 10 April 1912. The crossing ended ignominiously four and a half days later, after a collision with an iceberg. Some 1,496 lives were lost in the disaster, the worst in maritime history up to that point. In the end, the Cunarder *Carpathia* rescued *Titanic*'s survivors. The sinking sent shockwaves through the industry. *Olympic* was removed from service and thoroughly overhauled, with additional lifeboats and watertight bulkheads being installed.

A series of photographs taken on Crossing No. 138 East, 8–14 May 1912, less than a month after the *Titanic* disaster.

Below: Captain Sir James Charles is clearly visible on the starboard Bridge wing, with two officers on his left and one to his right. Notice the missing stairs to the Boat Deck, and that the original canister-style vent remains in place at this late date. (Jim Kalafus Collection)

Right: The photographer has turned to face aft, overlooking the starboard Boat Deck. Extra lifeboats are apparent, but are in a very makeshift, temporary arrangement that would change a number of times through the years. (Jim Kalafus Collection)

Far right: A man walks beside the skylight cover for the First Class Lounge, while the photographer takes the photo from the vicinity of the No. 4 funnel looking forward. (Jim Kalafus Collection)

One of *Titanic*'s passengers, Benjamin Guggenheim, would have been alive if he had sailed as planned aboard *Lusitania*; however, when the Cunarder's April sailing had been cancelled so she could be drydocked for maintenance, Guggenheim transferred to *Titanic*. During the sinking, Guggenheim reportedly said of himself and his manservant, 'We've dressed up in our best and are prepared to go down as gentlemen', and they did.[53] Another passenger who would have sailed in First Class on *Titanic*'s maiden crossing had instead opted to go early on *Lusitania* before she was laid up: this was Guglielmo Marconi, founder of the Marconi Company, whose technology enabled a rescue of *Titanic*'s survivors. Because of his decision, Marconi was in New York in the days immediately following the disaster, meeting

53 *The New York Times*, 20 April 1912.

From the Sun Deck, aft, looking forward along the port Boat Deck. This photo shows that the new post-*Titanic* lifeboats include regular lifeboats and collapsibles, stored under the ship's original contingent of boats. (Jim Kalafus Collection)

Top: Standing on the Promenade Deck overlooking the fantail and stern. (Jim Kalafus Collection)

Above: Third Class passengers promenading on the Forecastle. In rough weather, this space was uninhabitable. (Jim Kalafus Collection)

Titanic's surviving wireless operator and participating in the American inquiry into her sinking.[54]

Lusitania and her sister had, fortuitously, been in service for over four years by then, building a reputation for safety and reliability. Even so, neither sister had enough lifeboats on board for their full passenger and crew complement, a vital deficiency on nearly all the liners on the Atlantic then, but one inescapably highlighted in the *Titanic* disaster. Cunard quickly placed more collapsibles and lifeboats aboard *Lusitania* so that she would have lifeboat capacity for all. She sailed from Liverpool for the first time after the *Titanic* disaster on 28 April 1912. The additional lifeboat installation was nothing if not hasty; up to the last moment before departure, crewmen could be seen fitting the new boats on the Boat Deck. *Olympic*'s first sailing from Southampton after the disaster had been cancelled because a large portion of her firemen had mutinied. Although *Lusitania* did not carry any of *Olympic*'s intended passengers from that cancelled trip, she did carry the mails that had been allotted to *Olympic*. On the Cunarder's return trip from New York, she also carried some of those who had survived the *Titanic* disaster, including Sir Cosmo and Lucy, Lady Duff Gordon. Cunard sent *Lusitania* on the longer southward passage

54 *Titanic: Death of a Dream*, Wyn Craig Wade, Penguin Books, London: 1986.

Lusitania sails eastbound to Europe at dusk in early 1914, even as the sun sets on the era that brought her to life. (HFX Studios)

to further ensure that she avoided any encounters with icebergs. Cunard also quickly instituted the position of Staff Captain to oversee shipboard activities, especially lifeboat drills. During later refits, extra midship davits were installed to handle *Lusitania*'s increased lifeboat numbers.

When *Olympic* was removed from service in late 1912 for a complete refit and overhaul, *Lusitania* and *Mauretania* were once again the unrivalled crack liners on the Atlantic trade. This did not last long, however; in June 1913, there came a new competitor, Hamburg-Amerika's *Imperator*. This monster three-funnelled ship was the first to exceed a length of 900ft, sporting a gross tonnage of 52,117, some 20,000 more than *Lusitania*. *Imperator* could carry almost 5,800 passengers and crew within her cavernous interiors. Her First Class appointments were widely considered the finest and most luxurious ever seen on an ocean liner. It wasn't all bad news for Cunard, however. *Imperator* proved to be a 'first rate hotel but a third rate ship', in the words of Hamburg-Amerika's Director, Albert Ballin. She was horribly top-heavy – earning the dubious nickname 'Limperator' – and rather prone to fire, as well. In her first year of service, she never satisfied Hamburg-Amerika's hopes. During a major overhaul between 1913 and 1914, drastic measures were taken to reduce top weight and increase stability. About 10ft was cut off the tops of each of her funnels; less grandiose, lighter-weight materials were also substituted for many of her First Class appointments and tons of cement were unceremoniously dumped into her double bottom.

In 1914, Hamburg-Amerika put a second leviathan into service: *Vaterland*. She was even larger than *Imperator*, and proved to be a much better-built ship, benefiting from greater experience building liners of her ilk. *Vaterland* was astoundingly spacious and luxurious, but the two Cunard sisters still had the advantage of speed.

On 30 May 1914, a new Cunard liner, which was to be paired up with *Lusitania* and *Mauretania*, began her maiden voyage: *Aquitania*. At 901ft 6in long and 97ft wide, she sported a gross tonnage of 45,647. Although not quite as large as the

Countess de Bertier and her young son Arnaud arriving in New York aboard *Lusitania* in January 1914, at the end of Crossing No. 167 West. They are standing on the starboard Boat Deck; just behind them are the doors leading into the Entrance. (Library of Congress, Prints & Photographs Division, Authors' Collection)

Actresses Violet Heming and Mabel Norton aboard *Lusitania*, probably in the spring or early summer 1914. *Lusitania* often carried stars of the theatre and musical world on her decks. (Library of Congress, Prints & Photographs Division, Authors' Collection)

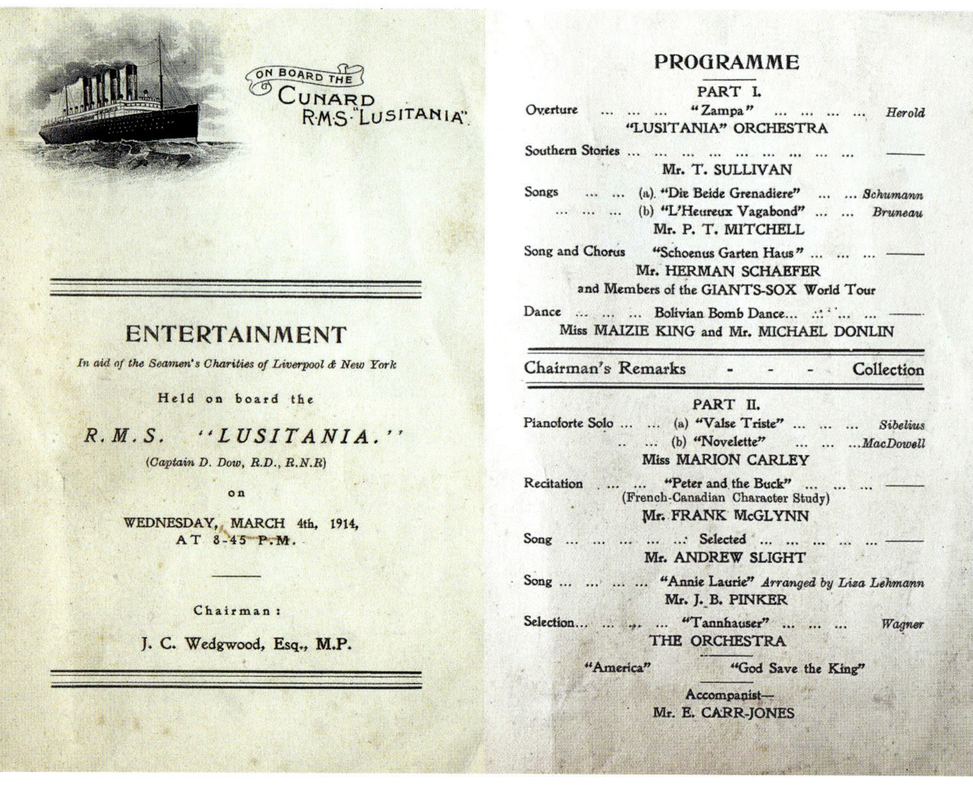

ON BOARD THE CUNARD R.M.S. "LUSITANIA".

ENTERTAINMENT

In aid of the Seamen's Charities of Liverpool & New York

Held on board the

R.M.S. "LUSITANIA."

(*Captain D. Dow, R.D., R.N.R*)

on

WEDNESDAY, MARCH 4th, 1914,
AT 8-45 P.M.

Chairman:

J. C. Wedgwood, Esq., M.P.

PROGRAMME

PART I.

Overture ... ... "Zampa" ... ... *Herold*
"LUSITANIA" ORCHESTRA

Southern Stories ... ... ——
Mr. T. SULLIVAN

Songs ... ... (a). "Die Beide Grenadiere" ... ... *Schumann*
... ... (b) "L'Heureux Vagabond" ... ... *Bruneau*
Mr. P. T. MITCHELL

Song and Chorus "Schoenus Garten Haus" ... ... ——
Mr. HERMAN SCHAEFER
and Members of the GIANTS-SOX World Tour

Dance ... ... Bolivian Bomb Dance... ... ... ——
Miss MAIZIE KING and Mr. MICHAEL DONLIN

Chairman's Remarks - - - Collection

PART II.

Pianoforte Solo ... ... (a) "Valse Triste" ... ... *Sibelius*
.. ... (b) "Novelette" ... ...*MacDowell*
Miss MARION CARLEY

Recitation ... ... "Peter and the Buck" ... ... ——
(French-Canadian Character Study)
Mr. FRANK McGLYNN

Song ... ... Selected ... ... ——
Mr. ANDREW SLIGHT

Song ... ... "Annie Laurie" *Arranged by Liza Lehmann*
Mr. J. B. PINKER

Selection... ... ... "Tannhauser" ... ... *Wagner*
THE ORCHESTRA

"America" "God Save the King"

Accompanist—
Mr. E. CARR-JONES

A concert programme from Crossing No. 171 West, March 1914. (Authors' Collection)

Lusitania arriving in New York at the conclusion of Crossing No. 171 West, 6 March 1914. (Library of Congress, Prints & Photographs Division, Authors' Collection)

This splendid starboard cutaway shows *Lusitania* as she appeared in the first half of 1914. (HFX Studios)

German sisters, she was longer than White Star's *Olympic* and *Britannic*, the latter of which was still unfinished; she was certainly competitive with the White Star and HAPAG giants. *Aquitania* was not designed to be as fast as *Lusitania* or *Mauretania*, but would instead make a more relaxed 23–4 knots. Having sacrificed that narrow margin of speed, she would give her passengers a thoroughly comfortable ride. Her hull was designed to ride over swells instead of dashing through them; inside, her First Class public rooms were lavishly and tastefully decorated to compete with the other first-rate liners plying the Atlantic. Placed under the command of Captain William Turner, and with her engines under the watchful eye of Chief Engineer Archibald Bryce, *Aquitania* was an instantaneous success. In her larger size and greater comfort, Cunard now had a well-balanced trio of ocean liners operating a weekly service between Liverpool and New York – the two fastest ships, and one of the most luxurious.

All of the excitement regarding the new *Aquitania* was still in the air when – quite suddenly and with very little warning – the whole of European civilisation collapsed. Thus began a cataclysm that quickly engulfed the entire world, ending the era that had given rise to *Lusitania*.

***

In Volume Two, we will explore how *Lusitania* adapted to new challenges during the war, and how the great greyhound met her demise. In doing so, we will also shed new light on details of the sinking that have been lost to history since 1915, and we will also see the turbulence left in the wake of the fate of one of the greatest ocean liners that ever graced the North Atlantic.

APPENDIX 1

FIRST CLASS PASSENGER LIST, CROSSING NO. 1

A
Abbott, Frank
Abelson, Abe
Ackerman, Ernest R.
Ackerman, Nora L.
Adams, Winifred (*Maid*)
Akahoshi, Fetsuma
Alexander, Kenneth D.
Alexander, Ralph
Anderson, Mary
Antoine, Gustave
Arnott, Hermione Louise
Arnott, Neil
Atkins, William Hugh
Atkinson, Francis Herbert
Attay, James Buesford
Ausoategui, Sabas
Austin, Frank Edward (*Valet*)

B
Badell, C. (*Maid*)
Bailey, Anna M.
Bailey, Charles
Bailey, Geo H.
Bailey, Grace L.
Bailey, Mrs Geo H.
Baird, E.
Balfour, Josephine Marie
Balfour, Robert
Bansman, Thomas F.
Barber, Ohio C.
Barnett, William J.
Bateman, James
Baxter, Blanche
Beeton, Mayson M.
Bell, James Wood
Bell, Thomas J.
Benn, Harrisire
Bentley, Wm John
Bevan, Anna
Bevan, Arthur
Bierce, Ellen D.
Bierce, Wm W.
Black, Frances
Blood, Aelish A.
Blood, Edward
Blood, Florie E.
Blood, Mary H.
Blood, Nancy
Blount, Hilda M.
Boron, Adelaide
Boron, Constance E.
Boron, Marjorie
Boston, Charles Fred
Bovee, Arthur Gibbon
Boyce, Cramer R.
Boyce, Hollie W.
Boyd, Hugh
Boyle, Frank Whelace
Branson, William George (*Valet*)
Bromwick, Francis Henry
Brown, Kirk
Brown, Peter Stuart
Browne, Lowell Huntington
Browne, Muriel Gore
Bulland, George Edwin
Butler, Fannie Marie
Butler, John William
Butsen, Henry H. Van
Butsen, Richard E.G. Van

C
Campbell, Hon. Cecil Arthur
Capewell, George Joseph
Carson, Charles George
Carteu de Marchieium, Mr G. (*Secretary*)
Cary, F. Stanley
Chafman, Clarence C.
Chafman, Evelyn S.
Charlesworth, A.H.
Charlesworth, Miss Eleanor
Charlesworth, Miss Mary
Charlesworth, Mrs A.H.
Chase, Frances Gertrude
Chetwynd, Amelie Mary
Child, Calvins
Chivers, Cedric
Christian, Rose
Christmas, Rose (*Nurse*)
Clark, Charles W.
Clark, Ernest L.
Clark, George
Clark, Jacobin Smith
Clark, Melville
Clark, Mrs Ernest L.
Clark, Ronald B.
Clarkson, Samuel
Clendenning, John
Clendenning, John J.
Cockshutt, Euiento Lisien
Colley, William H.
Collie, Alexander
Cooper, Huntting
Costijan, Lillian
Costijan, Miss Margaret
Costijan, Mrs Margaret
Crane, Elizabeth T.
Crane, Jas A.
Crane, John B.
Crane, Mrs
Crater, George E.
Croker, Elizabeth
Croker, Miss Ethel
Croker, Richard
Cunard, Ernest Haleburton
Cunningham, William C.
Cushner, Mrs. Samuel
Cushner, Samuel

D
Dahl, Rudolph
Davidson, Margaret B.
Davies, Arthur Henry
Davies, Earl B.
De Haven, Alexander Henry
De Haven, Mrs Alexander Henry
de Normandie, James
Dean, Henry Samuel
Denehy, Miss M.
Dilnot, Frank
Dixon, William
Dodge, Ethel
Dodge, Matthew H.
Doherty, Henry L.
Doherty, John P.
Dolan, Henry Gale
Donovan, John F.
Donovan, Robert F.
Doyle, Edward H.
Doyle, Mary E.
Doyle, Sarah J.
Drake, William
Dresser, Grace Forbush
Dresser, Miss De la Roy
Dresser, Miss Sue F.

Dresser, Mrs De la Roy
Dribbon, Henry J.
Dunbar, James R.
Dunbar, Lucilla Harris
Dunlap, Mrs
Dunlap, Robert
Dunmore, Gertrude

E
Edgar, Catherine Beatrice
Edgar, Charlotte
Edgar, Edward Mackay
Edgar, Ethel Brocadale
Edgar, John Fraser Mackay
Edgar, Mrs Edward Mackay
Edwards, Alury H.B.
Ettlen, Marie (*Maid*)

F
Fesler, James K.
Fibush, Aron
Flaitz, Albert
Forbes, George D.
Fow, Jenine R.
Fowlie, Miss Coutts
Fraser, Charles Lacklan
Freeman, Albert
Freeman, Fannie
Frick, Lena Jeffres
Frick, Master Hunter Jeffres
Frick, Mrs Hunter W.
Frost, Edwin P.
Frost, Mary C.
Frost, Ramelle M.K.
Furman, Mrs P.F.

G
Gadsdue, Arthur
Gallard, Joseph Adams
Gibb, David Eric
Giles, Anne (*Maid*)
Gill, Archie
Gillingham, Edward
Gillingham, Elsie
Glove, Chas F.
Goelet, Mrs Robert
Goelet, Robert
Goldmann, Robert
Graham, Mildred P.
Gray, Edith
Gray, James C.
Gray, Jessie C.
Green, George Harry
Grieve, Richard A.
Grimley, H.G.
Grogan, Bridget

H
Hands, Aaron A.
Hands, Anna S.
Hansen, Jos O.
Harde, Herbert Spencer
Harde, Mrs Herbert Spencer
Harriman, Herbert M.
Hay, Jennie B.
Hay, Louis C.
Hay, Wellington B.
Heath, Edward
Heber, Juliet E.
Hersey, Mrs Dudley H.
Higgenson, Frances Lee J.
Higgenson, Francis L.
Higgenson, Mrs Hetty
Hill, Edith
Hitchcock, Alfred Edwin
Hobbs, Walter Edward
Hodgeman, Lucy H.
Hodges, John S. Bach
Holt, Philip D.
Holt, William R.
Homewood, Ernest
Homewood, William Thos.
Hood, Joseph
Hook, Thos G.
Hope, Robert (*Valet*)
Hormis, Geo
Hubbard, Hermon Milton
Hubble, Maria
Hunt, William Floyd

J
James, Frederick (*Valet*)
Jean, Isaac
Jefferies, Jenny
Jefferies, Robert
Jeffries, Thomas
Jenny, William George
Jockura, William (*Valet*)
Joerinsen, Gertrude L.
Jones, Ernest Rae
Jones, George Albert
Jones, Lewis J.
Jones, Simeon
Jones, Wellington D.
Jonssen, Carl

K
Kellen, William Vail
Kelley, E.B.
Kelly, Mrs Elza
Kiernan, Stephen M.
Knight, Charles Y.
Knowles, John Thomas
Knowles, Mrs J.T.
Kowaki, Geryiro

L
Lacompte, Francis D.
Lajan, Hugh
Latham, Eleanor
Lever, Marty Edythe
Lever, Samuel H.
Leyand, Helen Dora
Leyland, Christopher Digby
Leyland, Christopher John
Leyland, Dorothy
Leyland, Helen Dora
Liberfierie, Marie (*Maid*)
Lindsay, William J.
Litchfield, Edward H.
Little, Jan Piteairie
Loel, Herman
Lomas, Frank E.
Lovegrove, Mrs Thomas G.
Lovegrove, Thomas G.
Low, Eliza
Low, John C.
Lowry, John A.
Ludlow, Thomas (*Valet*)
Luke, Ellen
Luke, Elsie Ellen
Luke, William Joseph

M
Mabson, Richard R.
MacBozis, Mrs
MacBozis, S.P. Rand
McCabe, Joseph L.
McCargo, Grant
McCargo, Mary
McCormick, Cyrus H.
McCormick, Gordon
McFadden, Geo H.
McFadden, Mrs Geo H.
Macfee, Anne (*Nurse*)
MacHugh, Robert Joseph
Mackay, Alexander
MacKenzie, George Richard
Mackey, Frank J.
Mackey, Mrs Frank J.
McMormick, Cyrus H.
McNiell, Daniel
Magnus, Anna
Mahan, Charlotte
Majima, Tomoyoshi
Mallach, John
Mallory, Albert
Markanian, Bedros H.
Marsh, Wilber W.
Mason, Alfred Darius
Masuda, Nobuyo
Masuda, Takashi
Mayhew, Cornelia
Maynard, Frank H.
Maynard, Mrs F.
Mendes, Edyard Pereica
Mendes, Octaviano
Millar, Charles
Millar, Eleanor
Millar, Florence
Millar, Gertrude
Moore, George F.
Moore, Samuel Wallace
Morrill, John F.
Morrison, D.L.
Morrissey, Daniel A.
Murphy, Dorothy
Murphy, Eugene
Murray, Lady Victoria
Myers, George
Myers, Mrs George

N
Nicholls, Joseph
Nielson, John Frederick
North, Hugh M. James
Norton, E. Hope
Norton, Lily M.

O
Ogilvie, Thomas
Onterbudge, Cyril N.
Owen, Alice
Owen, Samuel

P
Page, Howard
Paine, George F.D.
Paine, Margaret Elizabeth
Palmer, Mrs Potter
Palmer, Potter
Patten, Alexander
Patten, Edith B.
Patton, Mrs Alexander E.
Payne, George
Payne, Harriet
Payne, John
Payne, Rosa
Payne, Thomas
Peabody, George
Peter, John
Phillips, James Alexander
Pinder, Muriel
Plank, Milton H.
Plank, Mrs Mary

R
Ralston, Kate J.
Ralston, William Johns
Ralston, Wm J.
Rankin, Alexander
Rankin, Mrs Alexander
Reeves, Mary T.
Reeves, Samuel J.
Reidel, Anna
Reitleuger, Alexander
Reynolds, Mrs. Charles G.
Rice, Isaac L.
Rich, Clayton E.
Rich, Mrs J.
Roberts, Eileen
Ross, James
Rudd, Anthony George
Rudd, Anthory G.

S
Sandermann, David Thomas
Saunders, Elva
Saunders, Minerva
Sawyer, Samuel Woodson
Seccombe, Elizabeth
Sewell, Maria
Sewell, Marion (*Maid*)
Shavin, John H.
Shaw, Charles G.
Shaw, Charles G. Jessie
Shaw, Frank D. 2nd
Shaw, Frank D., Esq.
Sheedy, Mary
Shirra, Charles
Shirra, Edith Mercedes
Shirra, Mrs Charles
Shraker, Margurite
Sidebottom, Mrs William
Sidebottom, William
Simpson, Gertrude
Simpson, William
Sims, Harold Haig
Smith, Grace M.
Smith, Harold H.
Smith, James Nicoll
Somerset, Mrs Thomas
Somerset, Thomas
Soule, Beach C.
Soule, Carry J.
Soule, Ernest B.
Soule, Henry B.
Soule, Mrs E. Bacon
Stewart, Louis
Still, Andrew
Stimson, Miss Cordelia
Stimson, Mrs George W.
Stone, Henry A.
Stuart, Horatio
Sullivan, James
Suter, Hermann
Sutherland, George

T
Tabbersfield, Gerald
Tabbersfield, Kube Jos.
Tabbersfield, Olga
Tabbersfield, Percival
Tanakaa, Keishui
Tayler, Edith
Taylor, Arthur James
Taylor, Geo S.
Taylor, Janet (*Maid*)
Thompson, Ralph C.
Thompson, William Payne
Thomson, Benjamin Thomas
Thomson, Robert Bruce
Thorn, Charles E.
Thorn, Mrs Charles E.
Tiedmann, Tuder J.A.
Travers, Wm Richard
Triscott, Emily
Trumbull, Gertrude
Trumbull, Walter H.
Turner, Harold Rupert
Tweedle, Robert

V
Vallee, Sarah (*House Keeper*)
Van Liew, Harry
Verdon, Frederick A.
Verdon, Mr F.A.
Vogel, Martin

W
Wallbrook, Henry Mackinius
Walls, Annie
Walls, John
Walls, Reina
Ward, John
Ward, Mrs
Ward, Mrs Cordelia E.
Ward, Reginald H.
Warren, Julia P.
Warren, Louise
Warren, Moses A.
Watts, Ruth Budd
Webster, William
Weir, William
Welch, George (*Valet*)
Wellman, Frederick S.
Wellman, Julien A.
Wellman, Samuel T.
Wells, Harry W.
Wells, Sydney Clarke
Wenbury, Frank (*Valet*)
Werthinia, Jules
Wheeler, Arthur
Whippell, Sherman L.
Whitaker Gribben, Robert
White, Archibald S.
White, Mrs Archibald S.
Whittaker, Christopher Joseph
Whyham, Walter
Wigram, W.G.
Williams, Frank B.
Williams, Mrs F.B.
Williams, Richard
Wilson, Luke J.
Winterbotham, Lindsey Dillon
Wiseman, Sir George Eden
Wishart, George
Witherspoon, Preston
Wolfenden, Chas F.
Wood, Emersen
Worthington, H. Edward
Worthington, Henry E.
Wotton, Henry
Wulff, Deller Friedrich
Wylie, Herbert
Wylie, Mrs Herbert

Y
Young, Wm H.

The cover for *Lusitania*'s maiden voyage First Class passenger list. (Günter Bäbler Collection)

APPENDIX 2

VOYAGES AND CROSSINGS RMS *LUSITANIA*

Voyage no.:	Crossing no.:	Departure port and date (dd/mm/yyyy):	Intermediary port of call (dd/mm/yyyy):	Arrival port and date (dd/mm/yyyy):
1	1 West	Liverpool: Sat, 07/09/1907	Queenstown: Sun, 08/09/1907	New York: Fri, 13/09/1907
	2 East	New York: Sat, 21/09/1907	Queenstown: Fri, 27/09/1907	Liverpool: Sat, 28/09/1907
2	3 West	Liverpool: Sat, 05/10/1907	Queenstown: Sun, 06/10/1907	New York: Fri, 11/10/1907
	4 East	New York: Sat, 19/10/1907	Queenstown: Thurs, 24/10/1907	Liverpool: Fri, 25/10/1907
3	5 West	Liverpool: Sat, 02/11/1907	Queenstown: Sun, 03/11/1907	New York: Fri, 08/11/1907
	6 East	New York: Sat, 16/11/1907	Queenstown: Fri, 21/11/1907	Liverpool: Sat, 22/11/1907
4	7 West	Liverpool: Sat 30/11/1907 (DELAYED)	Queenstown: Mon, 02/12/1907	New York: Sun, 08/12/1907
	8 East	New York: Sat, 14/12/1907	Queenstown: Fri, 20/12/1907	Liverpool: Fri, 20/12/1907
5	9 West	Liverpool: Sat, 28/12/1907	Queenstown: Sun, 29/12/1907	New York: Fri, 03/01/1908
	10 East	New York: Sat, 11/01/1908	Queenstown: Thurs, 16/01/1908	Liverpool: Fri, 17/01/1908
6	11 West	Liverpool: Sat, 25/01/1908	Queenstown: Sun, 26/01/1908	New York: Sat, 01/02/1908
	12 East	New York: Sat, 08/02/1908	Queenstown: Fri, 14/02/1908	Liverpool: Fri, 14/02/1908
7	13 West	Liverpool: Sat, 07/03/1908	Queenstown: Sun, 08/03/1908	New York: Fri, 13/03/1908
	14 East	New York: Sat, 21/03/1908	Queenstown: Thurs, 26/03/1908	Liverpool: Fri, 27/03/1908
8	15 West	Liverpool: Sat, 04/04/1908	Queenstown: Sun, 05/04/1908	New York: Fri, 10/04/1908
	16 East	New York: Wed, 15/04/1908	Queenstown: Tues, 21/04/1908	Liverpool: Tues, 21/04/1908
9	17 West	Liverpool: Sat, 25/04/1908	Queenstown: Sun, 26/04/1908	New York: Fri, 01/05/1908
	18 East	New York: Wed, 06/05/1908	Queenstown: Tues, 12/05/1908	Liverpool: Tues, 12/05/1908
10	19 West	Liverpool: Sun, 16/05/1908	Queenstown: Mon, 17/05/1908	New York: Fri, 22/05/1908
	20 East	New York: Wed, 27/05/1908	Queenstown: Tues, 02/06/1908	Liverpool: Tues, 03/06/1908
11	21 West	Liverpool: Sat, 06/06/1908	Queenstown: Sun, 07/06/1908	New York: Fri, 12/06/1908
	22 East	New York: Wed, 17/06/1908	Queenstown: Tues, 23/06/1908	Liverpool: Tues, 23/06/1908
12	23 West	Liverpool: Sat, 04/07/1908	Queenstown: Sun, 05/07/1908	New York: Fri, 10/07/1908

	24 East	New York: Wed, 15/07/1908	Queenstown: Mon, 20/07/1908	Liverpool: Tues, 21/07/1908
13	25 West	Liverpool: Sat, 25/07/1908	Queenstown: Sun, 26/07/1908	New York: Fri, 31/07/1908
	26 East	New York: Wed, 05/08/1908	Queenstown: Mon, 10/08/1908	Liverpool: Tues, 11/08/1908
14	27 West	Liverpool: Sat, 15/08/1908	Queenstown: Sun, 16/08/1908	New York: Thurs, 20/08/1908
	28 East	New York: Wed, 26/08/1908	Queenstown: Mon, 31/08/1908	Liverpool: Tues, 01/09/1908
15	29 West	Liverpool: Sat, 05/09/1908	Queenstown: Sun, 06/09/1908	New York: Fri, 11/09/1908
	30 East	New York: Wed, 16/09/1908	Queenstown: Tues, 22/09/1908	Liverpool: Tues, 22/09/1908
16	31 West	Liverpool: Sat, 03/10/1908	Queenstown: Sun, 04/10/1908	New York: Fri, 09/10/1908
	32 East	New York: Wed, 14/10/1908	Queenstown: Mon, 19/10/1908	Liverpool: Tues, 20/10/1908
17	33 West	Liverpool: Sat, 24/10/1908	Queenstown: Sun, 25/10/1908	New York: Fri, 30/10/1908
	34 East	New York: Wed, 04/11/1908	Queenstown: Mon, 09/11/1908	Liverpool: Wed, 10/11/1908
18	35 West	Liverpool: Sat, 14/11/1908	Queenstown: Sun, 15/11/1908	New York: Fri, 20/11/1908
	36 East	New York: Wed, 25/11/1908	Queenstown: Tues, 01/12/1908	Liverpool: Thurs, 03/12/1908
19	37 West	Liverpool: Sat, 05/12/1908	Queenstown: Sun, 06/12/1908	New York: Fri, 11/12/1908
	38 East	New York: Wed, 16/12/1908	Queenstown: Mon, 21/12/1908	Liverpool: Tues, 22/12/1908
20	39 West	Liverpool: Sat, 26/12/1908	Queenstown: Sun, 27/12/1908	New York: Sat, 02/01/1909
	40 East	New York: Wed, 06/01/1909	Queenstown: Mon, 11/01/1909	Liverpool: Tues, 12/01/1909
21	41 West	Liverpool: Sun, 07/02/1909	Queenstown: Mon, 08/02/1909	New York: Sun, 14/02/1909
	42 East	New York: Wed, 17/02/1909	Queenstown: Mon, 22/02/1909	Liverpool: Tues, 23/02/1909
22	43 West	Liverpool: Sat, 27/02/1909	Queenstown: Sun, 28/02/1909	New York: Fri, 06/03/1909
	44 East	New York: Wed, 10/03/1909	Queenstown: Mon, 15/03/1909	Liverpool: Tues, 16/03/1909
23	45 West	Liverpool: Sat, 20/03/1909	Queenstown: Sun, 21/03/1909	New York: Fri, 26/03/1909
	46 East	New York: Wed, 31/03/1909	Queenstown: Tues, 06/04/1909	Liverpool: Wed, 07/04/1909
24	47 West	Liverpool: Sat, 17/04/1909	Queenstown: Sun, 18/04/1909	New York: Fri, 23/04/1909
	48 East	New York: Wed, 28/04/1909	Queenstown: Mon, 03/05/1909	Liverpool: Tues, 04/05/1909
25	49 West	Liverpool: Sat, 08/05/1909	Queenstown: Sun, 09/05/1909	New York: Fri, 14/05/1909
	50 East	New York: Wed 19/05/1909	Queenstown: Mon, 24/05/1909	Liverpool: Tues, 25/05/1909
26	51 West	Liverpool: Sun, 30/05/1909	Queenstown: Mon, 31/05/1909	New York: Fri, 04/06/1909
	52 East	New York: Wed, 09/06/1909	Queenstown: Mon, 14/06/1909	Liverpool: Tues, 15/06/1909
27	53 West	Liverpool: Sat, 19/06/1909	Queenstown: Sun, 20/06/1909	New York: Fri, 25/06/1909
	54 East	New York: Wed, 30/06/1909	Queenstown: Mon, 05/07/1909	Liverpool: Tues, 06/07/1909
28	55 West	Liverpool: Sat, 17/07/1909	Queenstown: Sun, 18/07/1909	New York: Fri, 23/07/1909
	56 East	New York: Wed, 28/07/1909	Queenstown: Mon, 02/08/1909	Liverpool: Mon, 02/08/1909
29	57 West	Liverpool: Sat, 07/08/1909	Queenstown: Sun, 08/08/1909	New York: Fri, 13/08/1909
	58 East	New York: Wed, 18/08/1909	Queenstown: Mon, 23/08/1909	Liverpool: Tues, 24/08/1909
30	59 West	Liverpool: Sat, 28/08/1909	Queenstown: Sun, 29/08/1909	New York: Thurs, 02/09/1909

	60 East	New York: Wed, 08/09/1909	Queenstown: Mon, 13/09/1909 Fishguard, Wales: Mon, 13/09/1909	Liverpool: Mon, 13/09/1909
31	61 West	Liverpool: Sat, 18/09/1909	Queenstown: Sun, 19/09/1909	New York: Fri, 24/09/1909
	62 East	New York: Wed, 29/09/1909	Qtown/Fishguard: Mon, 04/10/1909	Liverpool: Mon, 04/10/1909
32	63 West	Liverpool: Sat, 16/10/1909	Queenstown: Sun, 17/10/1909	New York: Thurs, 21/10/1909
	64 East	New York: Wed, 27/10/1909	Qtown/Fishguard: Mon, 01/11/1909	Liverpool: Mon, 01/11/1909
33	65 West	Liverpool: Sat, 06/11/1909	Queenstown: Sun, 07/11/1909	New York: Thurs, 11/11/1909
	66 East	New York: Wed, 17/11/1909	Qtown/Fishguard: Mon, 22/11/1909	Liverpool: Tues, 23/11/1909
34	67 West	Liverpool: Sat, 27/11/1909	Queenstown: Sun, 28/11/1909	New York: Fri, 03/12/1909
	68 East	New York: Wed, 08/12/1909	Qtown/Fishguard: Mon, 13/12/1909	Liverpool: Tues, 14/12/1909
35	69 West	Liverpool: Sat, 18/12/1909	Queenstown: Sun, 19/12/1909	New York: Fri, 24/12/1909
	70 East	New York: Wed, 29/12/1909	Qtown/Fishguard: Mon, 03/01/1910	Liverpool: Tues, 04/01/1910
36	71 West	Liverpool: Sat, 08/01/1910	Queenstown: Sun, 09/01/1910	New York: Sat, 14/01/1910
	72 East	New York: Wed, 20/01/1910	Fishguard: CANCELLED	Liverpool: Tues, 26/01/1910
37	73 West	Liverpool: Sat, 26/02/1910	Queenstown: Sun, 27/02/1910	New York: Fri, 04/03/1910
	74 East	New York: Wed, 09/03/1910	Fishguard: Mon, 14/03/1910	Liverpool: Tues, 15/03/1910
38	75 West	Liverpool: Sat, 19/03/1910	Queenstown: Sun, 20/03/1910	New York: Fri, 25/03/1910
	76 East	New York: Wed, 30/03/1910	Fishguard: CANCELLED	Liverpool: Tues, 05/04/1910
39	77 West	Liverpool: Sat, 09/04/1910	Queenstown: Sun, 10/04/1910	New York: Fri, 15/04/1910
	78 East	New York: Wed, 10/04/1910	Fishguard: Mon, 25/04/1910	Liverpool: Tues, 26/04/1910
40	79 West	Liverpool: Sat, 07/05/1910	Queenstown: Sun, 08/05/1910	New York: Thurs, 12/05/1910
	80 East	New York: Wed, 18/05/1910	Fishguard: Mon, 23/05/1910	Liverpool: Wed, 25/05/1910
41	81 West	Liverpool: Sat, 28/05/1910	Queenstown: Sun, 29/05/1910	New York: Thurs, 02/06/1910
	82 East	New York: Wed, 08/06/1910	Fishguard: Mon, 13/06/1910	Liverpool: Tues, 14/06/1910
42	83 West	Liverpool: Sat, 18/06/1910	Queenstown: Sun, 19/06/1910	New York: Thurs, 23/06/1910
	84 East	New York: Wed, 29/06/1910	Fishguard: Mon, 04/07/1910	Liverpool, Tues, 05/07/1910
43	85 West	Liverpool: Sat, 09/07/1910	Queenstown: Sun, 10/07/1910	New York: Thursday, 14/07/1910
	86 East	New York: Wed, 20/07/1910	Fishguard: Mon, 25/07/1910	Liverpool: Tues, 26/07/1910
44	87 West	Liverpool: Sat, 06/08/1910	Queenstown: Sun, 07/08/1910	New York: Thurs, 11/08/1910
	88 East	New York: Wed, 17/08/1910	Fishguard: Mon, 22/08/1910	Liverpool: Tues, 23/08/1910
45	89 West	Liverpool: Sat, 27/08/1910	Queenstown: Sun, 28/08/1910	New York: Thurs, 01/09/1910
	90 East	New York: Wed, 07/09/1910	Fishguard: Mon, 12/09/1910	Liverpool: Mon, 12/09/1910
46	91 West	Liverpool: Sat, 17/09/1910	Queenstown: Sun, 18/09/1910	New York: Thurs, 22/09/1910
	92 East	New York: Wed, 28/09/1910	Fishguard: Mon, 03/10/1910	Liverpool: Tues, 04/10/1910
47	93 West	Liverpool: Sat, 08/10/1910	Queenstown: Sun, 09/10/1910	New York: Thurs, 13/10/1910
	94 East	New York: Wed, 19/10/1910	Fishguard: Mon, 24/10/1910	Liverpool, Tues, 25/10/1910

48	95 West	Liverpool: Sat, 05/11/1910	Queenstown: Sun, 06/11/1910	New York: Thurs, 10/11/1910
	96 East	New York: Wed, 16/11/1910	Fishguard: Mon, 21/11/1910	Liverpool: Tues, 22/11/1910
49	97 West	Liverpool: Saturday, 17/12/1910	Queenstown: Sun, 18/12/1910	New York: Fri, 23/12/1910
	98 East	New York: Wed, 28/12/1910	Fishguard: CANCELLED 02/01/1911	Liverpool: Tues, 03/01/1911
50	99 West	Liverpool: Sat, 07/01/1911	Queenstown: Sun, 08/01/1911	New York: Thurs, 12/01/1911
	100 East	New York: Wed, 18/01/1911	Fishguard: Mon, 23/01/1911	Liverpool, Tues, 24/01/1911
51	101 West	Liverpool: Sat, 28/01/1911	Queenstown: Sun, 29/01/1911	New York: Fri, 03/02/1911
	102 East	New York: Wed, 08/02/1911	Fishguard: Mon, 13/02/1911	Liverpool: Tues, 14/02/1911
52	103 West	Liverpool: Sat, 18/02/1911	Queenstown: Sun, 19/02/1911	New York: Fri, 24/02/1911
	104 East	New York: Wed, 01/03/1911	Fishguard: Mon, 06/03/1911	Liverpool, Tues, 07/03/1911
53	105 West	Liverpool: Sat, 11/03/1911	Queenstown: Sun, 12/03/1911	New York: Thurs, 16/03/1911
	106 East	New York: Wed, 22/03/1911	Fishguard: Mon, 27/03/1911	Liverpool: Tues, 28/03/1911
54	107 West	Liverpool: Sat, 08/04/1911	Queenstown: Sun, 09/04/1911	New York: Thurs, 15/04/1911
	108 East	New York: Wed, 19/04/1911	Fishguard: Mon, 24/04/1911	Liverpool: Tues, 25/04/1911
55	109 West	Liverpool: Sat, 29/04/1911	Queenstown: Sun, 30/04/1911	New York: Thurs, 04/05/1911
	110 East	New York: Wed, 10/05/1911	Fishguard: Mon, 15/05/1911	Liverpool: Tues, 16/05/1911
56	111 West	Liverpool: Sat, 20/05/1911	Queenstown: Sun, 21/05/1911	New York: Thurs, 25/05/1911
	112 East	New York: Wed, 31/05/1911	Fishguard: Mon, 05/06/1911	Liverpool, Tues, 06/06/1911
57	113 West	Liverpool: Sat, 10/06/1911	Queenstown: Sun, 11/06/1911	New York: Thurs, 15/06/1911
	114 East	New York: Wed, 21/06/1911	Fishguard: Mon, 26/06/1911	Liverpool: Tues, 27/06/1911
58	115 West	Liverpool: Sat, 08/07/1911	Queenstown: Sun, 09/07/1911	New York: Thurs, 13/07/1911
	116 East	New York: Wed, 19/07/1911	Fishguard: Mon, 24/07/1911	Liverpool: Tues, 25/07/1911
59	117 West	Liverpool: Sat, 29/07/1911	Queenstown: Sun, 30/07/1911	New York: Thurs, 03/08/1911
	118 East	New York: Wed, 09/08/1911	Fishguard: Mon, 14/08/1911	Liverpool: Tues, 15/08/1911
60	119 West	Liverpool: Mon, 28/08/1911	Queenstown: Tues, 28/08/1911	New York: Sat, 02/09/1911
	120 East	New York: Sun, 03/09/1911	Fishguard: CANCELLED	Liverpool: Sat, 09/09/1911
61	121 West	Liverpool: Mon, 11/09/1911	Queenstown: Tues, 12/09/1911	New York: Sat, 17/09/1911
	122 East	New York: Wed, 20/09/1911	Fishguard: Mon, 25/09/1911	Liverpool: Tues, 26/09/1911
62	123 West	Liverpool: Sat, 07/10/1911	Queenstown: Sun, 08/10/1911	New York: Thurs, 12/10/1911
	124 East	New York: Wed, 18/10/1911	Fishguard: Mon, 23/10/1911	Liverpool: Tues, 24/10/1911
63	125 West	Liverpool: Sat, 28/10/1911	Queenstown: Sun, 29/10/1911	New York: Fri, 02/11/1911
	126 East	New York: Wed, 08/11/1911	Fishguard: Mon, 13/11/1911	Liverpool: Tues, 14/11/1911
64	127 West	Liverpool: Sat, 18/11/1911	Queenstown: Sun, 19/11/1911	New York: Fri, 24/11/1911
	128 East	New York: Wed, 29/11/1911	Fishguard: Mon, 04/12/1911	Liverpool: Tues, 05/12/1911
65	129 West	Liverpool: Sat, 08/12/1911	Queenstown: Sun, 09/12/1911	New York: Fri, 15/12/1911
	130 East	New York: Sat, 16/12/1911	Fishguard: Fri, 22/12/1911	Liverpool: Sat, 23/12/1911

66	131 West	Liverpool: Sat, 30/12/1911	Queenstown: Sun, 31/12/1911	New York: Fri, 05/01/1912
	132 East	New York: Wed, 10/01/1912	Fishguard: Mon, 15/01/1912	Liverpool: Tues, 16/01/1912
67	133 West	Liverpool: Sat, 17/02/1912	Queenstown: Sun, 18/02/1912	New York: Sat, 24/02/1912
	134 East	New York: Wed, 28/02/1912	Fishguard: CANCELLED 04/03/1912	Liverpool: Tues, 05/03/1912
68	135 West	Liverpool: Sat, 09/03/1912	Queenstown: Sun, 10/03/1912	New York: Fri, 15/03/1912
	136 East	New York: Wed, 20/03/1912	Fishguard: Mon, 25/03/1912	Liverpool: Tues, 26/03/1912
69	137 West	Liverpool: Sat, 27/04/1912	Queenstown: Sun, 28/04/1912	New York: Fri, 03/05/1912
	138 East	New York: Wed, 08/05/1912	Fishguard: Mon, 13/05/1912	Liverpool: Tues 14/05/1912
70	139 West	Liverpool: Sat, 18/05/1912	Queenstown: Sun, 19/05/1912	New York: Fri, 24/05/1912
	140 East	New York: Wed, 29/05/1912	Fishguard: Mon, 03/06/1912	Liverpool: Tues, 04/06/1912
71	141 West	Liverpool: Sat, 08/06/1912	Queenstown: Sun, 09/06/1912	New York: Sat, 15/06/1912
	142 East	New York: Wed, 19/06/1912	Fishguard: Tues, 25/06/1912	Liverpool: Tues, 25/06/1912
72	143 West	Liverpool: Sat, 06/07/1912	Queenstown: Sun, 07/07/1912	New York: Fri, 12/07/1912
	144 East	New York: Tues, 16/07/1912	Fishguard: Mon, 22/07/1912	Liverpool: Mon, 22/07/1912
73	145 West	Liverpool: Sat, 27/07/1912	Queenstown: Sun, 28/07/1912	New York: Sat, 03/08/1912
	146 East	New York: Wed, 07/08/1912	Fishguard: Tues, 13/08/1912	Liverpool: Tues, 13/08/1912
74	147 West	Liverpool: Sat, 17/08/1912	Queenstown: Sun, 18/08/1912	New York: Fri, 23/08/1912
	148 East	New York: Wed, 28/08/1912	Fishguard: Mon, 02/09/1912	Liverpool: Tues, 03/09/1912
75	149 West	Liverpool: Sat, 07/09/1912	Queenstown: Sun, 08/09/1912	New York: Fri, 13/09/1912
	150 East	New York: Wed, 18/09/1912	Fishguard: Mon, 23/09/1912	Liverpool: Tues, 24/09/1912
76	151 West	Liverpool: Sat, 28/09/1912	Queenstown: Sat, 28/09/1912	New York: Fri, 04/10/1912
	152 East	New York: Wed, 09/10/1912	Fishguard: Tues, 15/10/1912	Liverpool: Tues, 15/10/1912
77	153 West	Liverpool: Sat, 14/12/1912	Queenstown: Sun, 15/12/1912	New York: Sat, 21/12/1912
	154 East	New York: Tues, 24/12/1912	Fishguard: Mon, 30/12/1912	Liverpool: Tues, 31/12/1912
78	155 West	Liverpool: Sat, 23/08/1913	Queenstown: Sun, 24/08/1913	New York: Sat, 30/08/1913
	156 East	New York: Wed, 03/09/1913	Fishguard: Mon, 08/09/1913	Liverpool: Mon, 08/09/1913
79	157 West	Liverpool: Sat, 13/09/1913	Queenstown: Sun, 14/09/1913	New York: Thurs, 18/09/1913
	158 East	New York: Wed, 24/09/1913	Fishguard: Mon, 29/09/1913	Liverpool: Mon, 29/09/1913
80	159 West	Liverpool: Sat, 04/10/1913	Queenstown: Sun, 05/10/1913	New York: Fri, 10/10/1913
	160 East	New York: Wed, 15/10/1913	Fishguard: Mon, 20/10/1913	Liverpool: Tues, 21/10/1913
81	161 West	Liverpool: Sat, 25/10/1913	Queenstown: Sun, 26/10/1913	New York: Fri, 31/10/1913
	162 East	New York: Wed, 05/11/1913	Fishguard: Mon, 10/11/1913	Liverpool: Mon, 10/11/1913
82	163 West	Liverpool: Sat, 22/11/1913	Queenstown: DID NOT CALL 24/11/1913	New York: Fri, 28/11/1913
	164 East	New York: Wed, 03/12/1913	Fishguard: Mon, 08/12/1913	Liverpool: Mon, 08/12/1913
83	165 West	Liverpool: Sat, 13/12/1913	Queenstown: Sun, 14/12/1913	New York: Fri, 19/12/1913
	166 East	New York: Wed, 24/12/1913	Fishguard: Mon, 29/12/1913	Liverpool: Tues, 30/12/1913

84	167 West	Liverpool: Sat, 03/01/1914	Queenstown: Sun, 04/01/1914	New York: Fri, 09/01/1914
	168 East	New York: Wed, 15/01/1914	Fishguard: Mon, 19/01/1914	Liverpool: Tues, 20/01/1914
85	169 West	Liverpool: Sat, 31/01/1914	Queenstown: CANCELLED Sun, 01/02/1914	New York: Fri, 06/02/1914
	170 East	New York: Wed, 11/02/1914	Fishguard: Tues, 17/02/1914	Liverpool: Tues, 17/02/1914
86	171 West	Liverpool: Sat, 28/02/1914		New York: Fri, 06/03/1914
	172 East	New York: Tues, 10/03/1914	Fishguard: Mon, 16/03/1914	Liverpool: Mon, 16/03/1914
87	173 West	Liverpool: Sat, 21/03/1914		New York: Fri, 27/03/1914
	174 East	New York: Tues, 31/03/1914	Fishguard: Mon, 06/04/1914	Liverpool: Mon, 06/04/1914
88	175 West	Liverpool: Sat, 11/04/1914		New York: Fri, 17/04/1914
	176 East	New York: Tues, 21/04/1914	Fishguard: Mon, 27/04/1914	Liverpool: Mon, 27/04/1914
89	177 West	Liverpool: Sat, 09/05/1914		New York: Fri, 15/05/1914
	178 East	New York: Tues, 19/05/1914	Fishguard: Mon, 25/05/1914	Liverpool: Mon, 25/05/1914
90	179 West	Liverpool: Sat, 13/06/1914		New York: Fri, 19/06/1914
	180 East	New York: Tues, 23/06/1914	Fishguard: Tues, 30/06/1914	Liverpool: Tues, 30/06/1914
91	181 West	Liverpool: Sat, 04/07/1914		New York: Fri, 10/07/1914
	182 East	New York: Tues, 14/07/1914	Fishguard: Mon, 20/07/1914	Liverpool: Mon, 20/07/1914
92	183 West	Liverpool: Sat, 25/07/1914		New York: Fri, 31/07/1914
	184 East	New York: Tues, 04/08/1914		Liverpool: Tues, 11/08/1914
93	185 West	Liverpool: Sat, 12/09/1914		New York: Thurs, 17/09/1914
	186 East	New York: Wed, 23/09/1914		Liverpool: Tues, 29/09/1914
94	187 West	Liverpool: Sat, 03/10/1914		New York: Fri, 09/10/1914
	188 East	New York: Wed, 14/10/1914		Liverpool: Mon, 19/10/1914
95	189 West	Liverpool: Sat, 24/10/1914		New York: Sat, 31/10/1914
	190 East	New York: Wed, 04/11/1914		Liverpool: Tues, 10/11/1914
96	191 West	Liverpool: Sat, 21/11/1914		New York: Fri, 27/11/1914
	192 East	New York: Sat, 05/12/1914		Liverpool: Fri, 11/12/1914
97	193 West	Liverpool: Wed, 16/12/1914		New York: Wed, 23/12/1914
	194 East	New York: Wed, 30/12/1914		Liverpool: Tues, 05/01/1915
98	195 West	Liverpool: Sat, 16/01/1915		New York: Sat, 23/01/1915
	196 East	New York: Sat, 30/01/1915		Liverpool: Sat, 06/02/1915
99	197 West	Liverpool: Sat, 13/02/1915		New York: Sat, 20/02/1915
	198 East	New York: Sat, 27/02/1915		Liverpool: Sat, 06/03/1915
100	199 West	Liverpool: Sat, 20/03/1915		New York: Sat, 27/03/1915
	200 East	New York: Sat, 03/04/1915		Liverpool: Sat, 10/04/1915
101	201 West	Liverpool: Sat, 17/04/1915		New York: Sat, 24/04/1915
	202 East	New York: Sat, 01/05/1915		–

ACKNOWLEDGEMENTS

Our entire team would like to thank the following people for their assistance in bringing this book to life. First and foremost, we want to thank our commissioning editor Amy Rigg for believing in the two-volume, full-colour concept for this unprecedented book about *Lusitania*. Next, a heartfelt 'thank you' to our animator, digital modeller, artist, friend and supporter Alex Moeller: without your help, we couldn't have accomplished this, and you hung in there with us right to the finish line.

Stuart Williamson, for writing a fabulous foreword for this volume, for contributing so much material to fill its pages, and for all of your help as we researched and recreated *Lusitania* virtually. Syler Beaucage, who shared so much material from his personal collection with us for this book and who also rushed new scans over to us just in time for inclusion with this volume; Mark Chirnside, who shared so much archival and research data with us; Ellie Moffat, our friend and Lead Curator of Maritime History, National Museums Liverpool; Günter Bäbler, who was always willing to drop everything and help dig up wonderful new treasures for both of these volumes; Eleanor Gillers, Head of Rights and Reproductions, New York Historical Society Museum and Library; Kay Kays of Amgueddfa Cymru – Museum Wales; Mark Warren, for proofreading the original manuscript of this book all those years ago; everyone who opened their personal photographic and illustrative collections for use in the various editions of this book; and to the countless people who have and who continue to support our endeavours to preserve and tell the history of the great Atlantic liners of the past, you have our lasting gratitude.

J. Kent Layton would like to thank: First and foremost, I want to thank my dear wife Tessa for her patience with me as we worked to complete this project. Not to be forgotten are my parents, David and Cherie, and sister, Elyse, who have supported me on my maritime adventures through my entire life. A special thank you also goes out to all of my friends near and far, and to the many readers who enjoy and support my work. Last but not least, I must thank all of my fellow co-authors who worked with me on this project and helped bring it to life. Thank you all, from the bottom of my heart!

Tad Fitch would like to thank: First, I'd like to thank my father, Jerry Fitch, for inspiring my interest in history and writing. A huge thank you is owed to Jackie, Jocelyn and the rest of my family and friends, for always supporting my research, writing and other projects. I cannot send enough praise to my co-authors and collaborators on this project. It has been a joy working with all of you, even during the difficult times. Last, but not least, I would like to thank all of our readers and supporters. You are why we work on these projects and what inspires us to push ahead.

Michael Poirier would like to thank: Firstly, I'd like to thank my co-authors of this series. Our work together has been a joy. I would like to thank my mother, Cheryl Grayko, who took me to libraries, book stores and to meet *Titanic* survivors. I'd also like to thank my friends and family who aren't interested in ships but who are interested in the work that goes into the research. Many researchers who've helped us deserve thanks – Cliff Barry, Mike Beatty, George Behe, Shelley Dziedzic, Charles Haas, Brian Hawley, Peter Kelly, Paul Latimer, Jim Kalafus, Ellie Moffat, David Olivera, Trevor Powell, Eric Sauder, Jennifer Sellitti, Craig Stringer, Vic Verlinden, Bill Wormstedt, Russ Willoughby, Geoff Whitfield, Brandon Whited, Stu Williamson. The *Titanic* International Society, which has published many of our *Lusitania* articles. Lastly, the wonderful relatives of *Lusitania's* last voyage who have assisted over the years. We've worked hard to preserve their memories through the RMS *Lusitania* Association of Relatives and have accomplished great things.

Tom Lynskey would like to thank: I'd like to give a special thank you to my father who inspired my first love of history and the sea, kicking off what's been a long and exciting journey. I'd also like to thank the many historians and authors I've been surrounded by and had the pleasure of working with in one way or another throughout the past fifteen years, keeping my interest in these stories alive. My co-authors deserve my thanks as well for bringing their seemingly endless knowledge of *Lusitania* to the table, seeing this project through to the end, and holding down the fort at the times when I needed to focus on other projects. Even though it was mentioned in the Acknowledgements section, I want to thank our friend, Alex, for pulling off the impossible and producing whatever graphics were needed for this book, especially as we came down to the wire. Thank you to my wife, Emma, for her constant support and encouragement, as well as for helping to enhance some imagery for this book. Finally, I owe a massive amount of gratitude to my viewers and supporters on YouTube, Patreon and other platforms, who have given me a purpose to keep going with my studies and research, for pushing us in our endeavour that eventually became *Lusitania: The Greyhound's Wake* and even for giving our whole team the confidence to develop this book project.

Levi Rourke would like to thank: I would like to thank my parents Dan and Jacqueline, who for so many years took me on a relentless search for books, models and videos on all the ships I loved. I appreciate all my friends and my family that have listened and encouraged me on this incredible journey. I would also like to thank all the historians and fellow teammates who have given me the privilege to live out my dream and keep these ships alive, in digital form, for people to enjoy for years to come.

Also, while we have been meticulous in our research, if errors have crept into the text, please let us know so that we can make alterations in future editions of this volume. Any mistakes in our research are ours alone, and should be attributed to no one else.

BIBLIOGRAPHY

ARCHIVAL MATERIAL

Mersey Inquiry into the *Titanic* disaster, 1912
National Archives & Records Administration (US)
Senate Inquiry into the *Titanic* disaster, 1912
Cunard Archives, University of Liverpool
The Saga of the RMS Lusitania: A Marine Forensic Analysis, The Society of Naval Architects and Marine Engineers (SNAME), 1998

BOOKS

The Scientific American Handbook of Travel, Munn & Co.: 1910
Bailey, Thomas and Ryan, Paul, *The Lusitania Disaster*, The Free Press, New York: 1975
Chirnside, Mark, *The Olympic Class Ships*, The History Press, Stroud: 2011
Fitch, Tad and Layton, J. Kent, *The Unseen Aquitania*, The History Press, Cheltenham: 2024
Fitch, Tad and Poirier, Michael, *Into the Danger Zone: The Lusitania, First Battle of the Atlantic, and Liners during the Great War*, Blurb, San Francisco: 2021
Gould, John, *Outward Bound*, 1905/1913
Hoehling, A. A. and Mary, *The Last Voyage of the Lusitania*, Holt, New York: 1956.
Johnston, Ian, *Ships for a Nation: John Brown & Company Clydebank*, Argyll Publishing, Scotland: 2008
Jordan, Humfrey, *Mauretania: Landfalls and Departures of Twenty Five Years*, Patrick Stephens, Wellingborough: 1988
King, Greg and Wilson, Penny, *Lusitania: Triumph, Tragedy, and the End of the Edwardian Age*, St Martin's Press, New York: 2015
Layton, J. Kent, *Conspiracies at Sea: Titanic and Lusitania*, Amberley, Stroud: 2016
Layton, J. Kent, *Lusitania: An Illustrated Biography*, Amberley, Stroud: 2015
Layton, J. Kent, *The Unseen Mauretania*, The History Press, Stroud: 2021
Petrova, Olga, *Butter With My Bread: The Memoirs of Olga Petrova*, Bobbs Merrill, Indianapolis: 1942
Roughead, William, *The Trial of Oscar Slater*, J. Day, New York: 1927
Sauder, Eric, *Lusitania: Triumph of the Edwardian Era* (details, check name)
Sauder, Eric, *The Unseen Lusitania: The Ship in Rare Illustrations*, The History Press, Stroud: 2015
Sauder, Eric, *RMS Lusitania: The Ship and Her Record*, Tempus, Stroud: 2005
Talbot, Frederick, *Steamship Conquest of the World*, William Heinemann Company, London: 1912
Thomas, Lowell, *Raiders of the Deep*, Doubleday Doran, New York: 1928
Tweedie, Mrs. Alec, *My Tablecloths: A Few Reminiscences*, Hutchinson & Co, London, 1916
Verlinden, Vic, *Lusitania: The Underwater Collection*, WDT: Belgium: 2022
Wade, Wyn Craig, *Titanic: Death of a Dream*, Penguin Books, London: 1986
Warren, Mark, *Ocean Liners of the Past vol 2*, Blue Riband Press, New York, 1997

PERIODICALS/NEWSPAPERS/JOURNALS

Aberdeen Press and Journal
Bath Chronicle
The Builder
The Commutator, journal of the Titanic Historical Society
The Daily Mirror
Dawson Daily News
Dundee Courier
Electrical Engineering
Electrical Review and Western Electrician
Engineering: An Illustrated Weekly Journal
Evening Express & Evening Mail
Exeter and Plymouth Gazette
The Metal Industry
The Morning Leader
Munsey's Magazine
New Outlook
The New York Sun
The New York Times
Poverty Bay Herald
The Sketch
The Sphere
Steamship and Other Power Vessels
Telephony
The Van Wert Daily Bulletin
The Washington Herald
Western Times

INDEX

Note: **bold** page references denote illustrations